CULTURE COLLECTIONS

The Specialists' Conference on Culture Collections was sponsored by the Canadian Committee on Culture Collections of Micro-organisms, an associate committee of the National Research Council of Canada.

CULTURE COLLECTIONS:
Perspectives and Problems

Proceedings of the
Specialists' Conference on
Culture Collections,
Ottawa, August 1962

Edited by S. M. MARTIN

Division of Applied Biology
National Research Council
Ottawa, Canada

UNIVERSITY OF TORONTO PRESS

Published and Printed in Canada by
UNIVERSITY OF TORONTO PRESS, 1963
All Rights Reserved
Reprinted in 2018
ISBN 978-1-4875-8143-5 (paper)

PREFACE

This, the first international Specialists' Conference on Culture Collections, was organized with several aims. It was hoped that a consideration of the role of culture collections in science and industry would place the importance of culture collections in true perspective for all to see. Secondly, through a consideration of the organization of collections, we hoped, not only to aid workers actively concerned with culture collections, but also to give a clear picture of the organizational problems to administrators. We also wished to give consideration to the fundamental and technical aspects of the preservation of micro-organisms and other cells, with special emphasis being placed on the maintenance of physiological, morphological and genetic characteristics. The speakers have striven to elucidate what is known of the subject matter, and also to draw attention to areas of ignorance, with the hope of stimulating research in these areas. Last, but not least, it was felt that, by bringing together specialists with varied backgrounds and interests, problems of mutual, but perhaps unrecognized, concern might be brought to light.

Some may feel that our aims have been too all-encompassing for such a short meeting. However, if the conference has done nothing more than to leave food for thought, it can be considered a success. If these proceedings can offer the same fare to the reader, they too will have served their purpose.

As secretary of the conference I extend my sincere thanks to all of those who took part in the program. Especial thanks are due Drs. N. E. Gibbons, I. J. McDonald and C. Quadling who aided so greatly in the planning and organization of the conference.

It has been a great privilege and a rich experience to have been associated with this conference. It is rewarding to know that the conference-at-large voted unanimously for the continuation of such conferences on an international basis.

Ottawa, Canada   S. M. Martin
August, 1962

FUTURE CULTURE COLLECTION CONFERENCES

Some of the delegates to this conference have inquired whether the work and interest started by the organizers of the conference could be continued by arranging for similar meetings in the future.

A group of interested people met informally last evening to discuss this question. The concensus of this ad hoc committee was that, considering the evident interest shown by workers in this field, the wide and basic applications of the work of culture collections and the opportunities for increasing international co-operation in the field, it would be appropriate that future conferences be held.

The group agreed that the organization of such conferences would be facilitated if they were held under the auspices of the International Association of Microbiological Societies. The recommended course of action was to request authorization of the Ottawa Conference to form an international committee whose responsibility it would be to organize future meetings of culture collection specialists. Such sanction being obtained, the committee could then apply to I. A. M. S. for recognition as a Section on Culture Collections.

Therefore, the ad hoc committee, which met last evening, agreed to present the following recommendations to this conference of specialists:
1) that the Conference recommend that future conferences on culture collections be held;
2) that the Conference recommend the formation of an International Committee on Culture Collections, such committee to apply for recognition as a Section of I. A. M. S. ;
3) that the membership of said Committee be approved as follows:

Prof. T. Asai	Prof. N. A. Krassilnikov
Miss A. L. van Beverwijk	Dr. S. M. Martin
Dr. W. A. Clark	Dr. V. B. D. Skerman
Dr. R. Donovick	Dr. K. J. Steel (acting permanent secretary)

The Committee will elect a chairman at a later date.

These recommendations were passed by unanimous vote of the Conference-at-Large.

W. A. Clark

Ottawa, Canada
28 August 1962.

CONTENTS

WELCOME TO DELEGATES

by W. H. Cook, Director
Division of Applied Biology
National Research Council
Ottawa, Canada

It is a pleasure for me to welcome this specialists' conference on culture collections to the National Research Council of Canada. It seems appropriate that you should meet in Ottawa as the average Canadian considers Ottawa the home of collections, particularly income taxes. Tourists often say it is a beautiful city, but the inhabitants sometimes describe its features in less complimentary terms. Politicians seem to consider Ottawa a desirable place. Scientists usually find the research institutes and laboratories of the Federal Government of greatest interest.

We are still a developing country but research has been done in such government departments as Agriculture, and Health and Welfare, for nearly a century. The National Research Council was formed in 1916 as a separate research agency charged with the responsibility for stimulating, supporting, and performing research without the limitation of departmental boundaries and responsibilities. During the first twelve years of its existence, this Council devoted its entire appropriation to the support of university research. This has been continued and today Canadian universities obtain about a third of their total research funds from the National Research Council.

In 1928 the Council opened its own laboratories, organized four research divisions and grew slowly until 1939 when it had a total staff of about 250. Today, the Council has 10 research divisions in its Ottawa laboratories, two regional laboratories, and a total staff of about 2,500. In the meantime the Defence Research Board, Atomic Energy of Canada, and the Medical Research Council are independent agencies that have grown out of activities once performed by N. R. C.

A similar growth in scientific activity has occurred elsewhere in Canada and throughout the world in the past two decades. Personally, I wonder if our means of communication and exchanging scientific information has kept pace with this growth and development. International congresses and even national scientific meetings have increased in number but continued as omnibus sessions and grown to a size that seems to have decreased rather than increased their efficiency.

Personally, I feel two changes in the organization of scientific meetings are required to modernize this phase of our communication. First, there should be a sharper definition and delineation of the subject matter to be covered at each international congress.

It is possible today to have a single paper on, say, some fundamental field of microbiology found acceptable for presentation at a half-dozen international congresses. Second, we need more specialists' meetings of this sort where those interested in a particular field have an opportunity to exchange views and detailed information. I think you must share this opinion as otherwise there would not be such a large attendance at this meeting immediately after an International Congress on Microbiology. I welcome you particularly for this reason, since it is always satisfying to have a group that shares similar views to one's own.

But specialization does not mean fragmentation, and I am alarmed by the trend toward fragmentation in the life sciences. Measured in terms of funds or scientific manpower, research in the physical sciences is much larger than research in the life sciences. Yet the physical sciences have retained much of their fundamental unity - mathematics, physics, chemistry and engineering are still taught primarily in university departments and faculties with these names.

By comparison, life science is being taught in an ever-increasing number of university departments, and there is an increasing clamor for more international unions in life science. This tendency has arisen for many diverse reasons, in part it may reflect our failure to organize sufficient specialists' meetings of this sort, and perhaps the taxonomic training of biologists has encouraged them to distinguish small differences. For instance, in your own field there are microbiologists, bacteriologists, zymologists, virologists, cytologists, serologists, immunologists etc. extending to a binomial nomenclature such as cellular biologists, electron microscopists, molecular biologists, biochemical geneticists etc., all of whom may work on micro-organisms. We seem to have named all the approaches to the subject but forgotten the subject itself. Recently personnel trained in physics and chemistry have undertaken research on biological material. This has been most helpful and productive, but has given rise to even more subdivisions. Basically, these people are biologists since they are advancing life science rather than physical science.

Culture collections have great practical significance in agriculture, industry and medicine. But it is much more than this, it involves some of the most fundamental studies on the nature of living processes. Preservation must limit growth and metabolism, maintain viability and also specific biological properties of the organism. This has been done by modifying the external environment, the approach to the new environment, and the rate at which the change is made. Much of the information is empirical, and little is known as to why a particular procedure that is successful with some organisms fails to preserve others.

In this work you will have some assistance from those who

use micro-organisms as "guinea pigs" for fundamental studies on living processes. Micro-organisms can often withstand freezing and other environmental extremes that are lethal to the multicellular organism with its complex organization. It also appears that the tolerance of environmental extremes decreases when the organization of the organism is simplified below some critical point. Phages are generally more fastidious than their hosts - I see some evidence of this in the reports you are to discuss.

In preserving even the simplest micro-organism, you are attempting to control the galaxy of conditions that constitute life itself. You modify some of the conditions in the external environment, and hope that this will adjust the constellation of conditions in the cytoplasm so as to suspend, but not destroy, the metabolic potential of the enzymes and RNA. Inside this again is another constellation, the genetic code, DNA and its postulated masks and messengers. Many environmental conditions, ranging from the presence of transforming principles to electromagnetic and ionizing radiations, are known to alter the genetic code. In general, the probability of mutations is greatest under environmental extremes - the conditions you must often impose for successful preservation. Much has been accomplished but there is much more to be done in these basic fields. One cannot chart a course through the unknown but I hope this meeting will provide opportunities for exchanging information that will prove to be helpful. Even an idea with which you may disagree can be useful if it stimulates new work.

These philosophical generalities from a biochemist who knows little about culture collections are akin to the froth on that end product of microbial metabolism called beer. I shall now let you proceed with more palatable fare. We are glad to have you with us and I hope that your sessions are both pleasant and profitable. If you think we can be of further help to you, individually or collectively, on any aspect of this meeting or visit, please ask about it - we may not be able to help, but we can try.

CHAIRMAN'S REMARKS

by N. E. Gibbons
National Research Council
Ottawa, Canada

Some of you may have been wondering how this conference came about and just what the Canadian Committee on Culture Collections of Micro-organisms is. So a brief outline of the background may be worthwhile.

The British Commonwealth Scientific Official Conference, held in 1946, recommended that a Commonwealth Organization be established to maintain collections of type cultures, and that a specialists conference be called to consider the question. It also suggested that a suitable time for such a conference was the occasion of the IV International Congress in Copenhagen. Consequently a Specialists Congress met in London in August 1947. On the recommendation of the Conference the "British Commonwealth Collections of Micro-organisms" was established and National Committees were set up in the larger commonwealth countries to implement its proposals. Thus the Canadian Committee was established in 1948, as an associate committee of the National Research Council. The main function of these national committees has been to prepare directories of existing culture collections, and to encourage the publication of catalogues of the larger collections.

At the meeting of the Permanent Committee of the British Commonwealth Collections of Micro-organisms, held after the Stockholm Congress, the suggestion was made that a symposium might be held after the Montreal Congress on the Operation of Culture Collections. Since one of the original recommendations of the 1946 BCSO Conference was that "the attention of the United Nations Organization should be drawn to the desirability of establishing an international body to extend coordination to the international field" - the organizers of this Conference felt that attendance should not be limited to commonwealth countries and we are pleased that today we have delegates representing 26 countries.

Perhaps I should make a brief digression to consider world coordination of culture collections. Of course there has been some coordination through the Centre de Collection de Type Microbiens at Lausanne and the suggestion of a World Catalogue is further evidence. Is there need for more? I might say that a small group is meeting this evening to consider the advisability of other Culture Collection Conferences and if so, under whose auspices.

When we began to think of a program, we had the temerity to think that some of the problems we had discussed in connection with our own small collection might be worth considering at a meeting such as this - the best way to operate a collection, the principles behind the various methods of preserving organisms, a critical assessment of the methods in common use. Correspondence with a limited number of workers in various parts of the world seemed to confirm our thoughts and so the program as you see it today. To get answers to these questions, we need the assistance of most of the disciplines and subdivisions mentioned by Dr. Cook. And we hope the result will not be further fragmentation but rather a bringing together of ideas so that we may gain a little better insight into what must be done to keep organisms with the same characteristics as when they were placed in the collection. We hope that in the papers and discussions of these two days we may get a few answers. If this is not possible, perhaps we can hope for a better delineation of the problems curators of culture collections face.

ROLE AND ORGANIZATION OF CULTURE COLLECTIONS

THE ROLE AND ORGANIZATION OF CULTURE COLLECTIONS

INTRODUCTORY REMARKS

by R. E. Buchanan
Iowa State University
Ames, U.S.A.

First, as members of this conference we should congratulate and thank Dr. N. E. Gibbons and Dr. S. M. Martin and their committee for their timely recognition of the need for an international conference of specialists concerned with culture collections of micro-organisms and on the eminently satisfactory development of the facilities and the programme.

Contacts, through many years, with culture collections of local, national and international sponsorships, with their governing boards and different institutional relationships, with the International Association of Microbiological Societies (IAMS) and with the International Union of Biological Societies have convinced me not only of the basic scientific importance of such collections but of their utilitarian significance as well. We are happy to be here.

After a careful perusal of the papers to be presented at this opening session, I have concluded that the only contribution I can make in introducing the speakers is to note three overall or general problems confronting all culture collections, items which may be otherwise overlooked.

1. The first of these problems relates to confusion as to the recognition of the nomenclatural heirarchies of names of taxa contained in the respective collections.

(a) Carefully to be distinguished are the type or neotype strains of the type species of each genus. One of the major scientific contributions to be made by a culture collection is the identification and designation of type or neotype strains for which suitable proposals have not been made previously.

(b) Also to be distinguished are the type or neotype strains of each species or subspecies for which types or neotypes have been designated. Participation of the staffs of culture collections is likewise much needed in the designation of type or neotype strains of all the species and subspecies in the several collections whose names are validly published and legitimate.

(c) Further, we need to distinguish additional named strains of species and subspecies of sufficient scientific, technical, analytical, genetic or industrial significance that differ in one or more characters from the type or neotype strains. Cultures of such strains should be designated as type or neotype cultures of infrasubspecific forms of the species.

  The nomenclatural goal of a culture collection should be the recognition of these three categories and the cataloguing of each strain under its correct name or designation.

  2. The second of these problems has to do with the fact that a culture collection may contain micro-organisms classified as belonging to the bacteria, to the plant kingdom or to the animal kingdom. The bacteriological code of nomenclature serves for the bacteria, the botanical code for the algae, the fungi and the myxomycetes and the zoological code for the protozoa. Unfortunately, the rules in the three codes relative to typification do not agree in some respects. Culture collections should protest the perpetuation of unnecessary differences in terminology.

  3. The third problem is the incorporation of suitable nomenclatural guidance that may well result from studies in numerical taxonomy. We may expect material assistance, particularly on the infrasubspecific level.

CULTURE COLLECTIONS, WHY AND WHEREFORE

by A. L. van Beverwijk
Centraalbureau voor Schimmelcultures
Baarn, Netherlands

It may, I think, be useful to start a two days' conference by outlining some basic ideas on the subject to be discussed. Almost every institute, where work is done with micro-organisms, maintains a smaller or larger collection, in order to have cultures available for research and to preserve organisms that have proved to be of interest. These micro-organisms may have been isolated during the course of the research carried out at that institute, or they may have been obtained from other collections. Concerning culture collections in general, some main points will be considered, viz.:
1.   The differentiation of the existing collections into specialized collections and service collections.
2.   Defining their scope and function.
3.   The responsibilities of service collections as depository and distribution centres.
4.   Culture collections as centres of active study; cooperation between specialized and service collections.
5.   The need of an identification service.
If, in the following, I should lay more emphasis on fungi than on the other micro-organisms, it is by no means, because I rank them any higher in importance, but simply because I prefer to talk from experience, and being a mycologist, I feel most at home with fungi.

Specialized and service collections

In a report on culture collections, written at the request of the Bureau of the International Council of Scientific Unions in 1950, Professor Kluyver differentiated the existing culture collections into collections of the specialist type and those of the service type. By specialist collections he understood the smaller or larger collections of cultures that are principally needed for the research done at the institute maintaining the collection. Generally it is only by personal favour of the director that cultures from these collections are made available to workers outside the institute. At the service collections, on the other hand, the micro-organisms are maintained with the purpose of serving microbiologists all over the world. They are internationally organized and maintain a wide range of micro-organisms. Service collections are: the American Type Culture Collection in Washington, the culture collection of the Commonwealth Mycological Institute in Kew, England, the Collection of the Laboratoire de Cryptogamie in Paris and the Centraalbureau voor Schimmelcultures in Baarn, Netherlands. Between these two types of collections there exist all kinds of

transitions. The larger collections of the specialist type often
serve as centres of distribution to research workers in their own
country. As such they have a regional scope. The British Commonwealth designates these collections with the prefix "national"
initial to the title. This differentiation may help to classify the
many culture collections and to define their various tasks. Starting with the specialized collections: they can be specialized as to
the subject studied, e.g. medical, phytopathological; or they can be
specialized as to the organisms maintained, such as a collection
of Leptospira, of Penicillia, and so on.

Specialized as to organism: Collections maintaining a taxonomically limited group of organisms are built up by authorities in that
group, they are often the base of a monograph, or other purely
scientific investigation. A few examples from the domain of the
fungi are: Wollenweber's Fusaria, Thom and Raper Penicillia and
Aspergilli, The Yeasts at the C.B.S. in Delft, the water-moulds of
Coker and Couch, the Pythium's of Middleton. It is not rarely that
the interest of the investigator shifts to some other subject and he
can no longer maintain the organisms he studied. Such collections
threaten to be as transient as human beings are, unless the very
valuable cultures of these well studied micro-organisms are
deposited in time in the more stable service collections.

Specialized as to subject: Collections that are built up on the research done at an institute are very numerous. On the whole their
activities tend towards applied scientific research. Concerning the
subject of investigation, they can be differentiated into:
- medical and veterinary collections,
- phytopathological and agricultural collections,
- collections of organisms important for fermentation,
- collections at university laboratories for teaching and purely
  scientific research (taxonomy, physiology, genetics).

These specialists collections are generally centres of very active
research, the organisms are handled by authorities in their field,
fresh strains being regularly isolated and added to the collection.
Some of the larger of these collections issue catalogues and have
cultures available for exchange, or for a fee. At the C.B.S. we
have about 40 catalogues and lists of cultures published after the
second World War. These are always a great help for locating cultures requested from the C.B.S.-collection, but not available at
the moment. The Permanent Committee of the British Commonwealth Collections of Micro-organisms issues General Directories,
listing the collections and the cultures maintained in the United
Kingdom with Ghana and Trinidad, in Australia, New Zealand,
Canada and India. The scope of these collections is indicated by
the institutes that maintain them. On the good functioning of a
specialist collection of micro-organisms may depend the health of

a country, or the yield of crops. The medical and veterinary institutes have their serological strains, and type organisms ready for
diagnosis in case of a threatening epidemic. At phytopathological
institutes the life cycles of plant parasites are studied, as an aid
for exactly timing the moment when measures have to be taken or
when a fungicide should be applied.

A relatively new specialism are the collections of fermentation organisms. They are the outcome of the rapidly expanding
importance of micro-organisms in industrial processes. The organisms maintained range from the old time baking and brewing
yeasts, to the modern interest in Actinomycetes for antibiotic research. The research departments of fermentation industries
generally have collections of their own, carefully guarding their
production organisms and strains that prove to have promising
biochemical properties. For obtaining cultures the fermentation
industries are either dependent on other culture collections or
they may make their own isolates. It is our experience that, for
the exact identification of isolates, industries often have to appeal
to taxonomists of the culture collections. Whereas there are very
many catalogues listing micro-organisms of medical and phytopathological interest, there are only a few catalogues published
concerning fermentation organisms e.g. the catalogues of the
National Collection of Industrial Bacteria in Aberdeen and that of
the Japanese Institute of Fermentation Technology in Osaka. The
Northern Utilization Branch of the United States Department of
Agriculture in Peoria does not issue a catalogue. The curator of
the collection, however, is most willing to supply cultures to other
investigators, as we can see from the many N.R.R.L. numbers
designating cultures from this laboratory in other culture catalogues and publications.

The service collections

The collections of the service type have a two-sided scope:
they have to preserve valuable species and strains, and to distribute subcultures of these to other workers all over the world. As
depository centres these collections have to ensure the continued
existence of micro-organisms that are of importance to science
and industry: type cultures and authentic cultures, assay organisms
and strains that have been object of research work and should be
maintained for further investigation. A recent extension to this
task is, that service collections are asked, to take strains in deposit, for which a patent application is pending, the depositers
requiring the guarantee that their cultures are not distributed
without their written permission.

We commonly use the term "Type Culture Collection" in a
very general way, e.g. The American Type Culture Collection, The
National Collection of Type Cultures. This does not mean to say

that all the cultures maintained are derived from Type material.
The name Type Culture Collection was in use, years before the
Botanical Code of Nomenclature introduced the term "type" mean-
ing "nomenclatural type", which is, the original material of the
new species which fixes the application of the name concerned. By
far the greater number of the Hyphomycetes (Fungi imperfecti)
now present in Type Culture Collections, were not cultivated at the
time they were described, hence no real "type cultures" of these
species exist. The loosely used term "type culture" just means:
typical culture, conforming to the original description. At the
Botanical Congress in Montreal in 1959 cultures of fungi were
officially recognized as potential type material. To mycologists
type cultures of fungi, in the sense of nomenclatural types, are a
point of deep concern. In the International Code of Botanical
Nomenclature 1961, Article 7, note 2, it reads: "A holotype is the
one specimen or other element used by the author or designated
by him as the nomenclatural type". The nature of type material is,
that it always is either part or all of the original material from
which the description was drawn up. In case of unicellular organisms, bacteria and yeasts, the part of a colony equals the whole,
and subcultures can still be considered as part of that which the
author designated as type. It may all be derived from one individual. Here I may refer to the five possible interpretations of the
term individual, suggested by Dr. Buchanan. (1) If, however, we
are concerned with morphologically more complicated organisms,
such as fungi, which are differentiated into mycelium and fructifi-
cation organs, often both sexual and asexual, one part of the individual no longer equals the whole. A culture derived from type
material, but developing only the asexual sporulation or only
sterile mycelium can no longer serve as "type".

Bacteria and yeasts are described mainly from their physiological characteristics, which, as a rule, keep well in culture.
Mycologists, however, know only too well the bitter experience
that, notwithstanding all attempts with appetizing media, an Asco-
mycete may loose the property of producing asci in prolonged culture. Then only the imperfect form, the asexual way of spore
production will remain, and even this may get lost. The culture
then has lost all the properties of being type material.

In Recommendation 7A of the Montreal Code it reads: "when
living material is designated as a type (for Bacteria and Fungi only)
appropriate parts should be immediately preserved". No indication, however, is given about the method of this preservation. If a
new species was growing on a solid substrate, the author, as a rule,
deposited dried type material in a recognized herbarium where it
continues well preserved. If however, a new mould is isolated
from air, water or soil, the description of the species is based on

morphological characteristics as observed in culture. In this case a real "type culture" does exist. Here the problem arises how to preserve this most valuable type material. At the Commonwealth Mycological Institute a method has been worked out, to dry Petri dish cultures and fix them in cardboard boxes of special design, so that the structures of the fungus in the thin agar film can be readily examined (2). The dried down cultures together with slides in separate boxes, are incorporated in the herbarium. In this way the morphological characteristics of the fungus continue observable, in case they should no longer be fully produced in culture. Besides, the living culture remains available for all further research. I would highly recommend this method to be applied for all "type cultures" of fungi immediately after they have been received from the author of the species. Lyophilization also proves a satisfactory method for preservation, but not all fungi can be lyophilized, and a fungus-taxonomist is scarcely eager to have to revive a lyophil pellet into a culture again, before being able to examine the species.

In our experience it is necessary to check all new accessions, before incorporating them into the collection. Cultures received from the author of the species generally breed no problem, although occasionally a species may seem to us not quite as new to science as the author himself considered it to be. Sometimes cultures received in exchange are contaminated, or the identification by the isolator proves to be incorrect. Therefore all cultures received have to be carefully examined, if possible, by a mycologist well versed in that particular group. This means that a collection should have a large staff of specialized microbiologists and mycologists. This ideal situation is hardly realizable.

The second and certainly not less important purpose of a service collection is, to make the cultures readily available to other scientists. As distribution centres they have a very extensive and elaborate task to fulfil. The chief responsibilities to which a distribution centre should conform as far as possible, are:

1. maintaining a great diversity of species and strains.
2. following up the taxonomic as well as the biochemical literature, so that new species and strains of industrial value may be requested for incorporation into the collection.
3. keeping the organisms in as good a condition as possible, in case of fungi: to search for such circumstances of medium, light and temperature, as to maintain sound sporulation.
4. keeping detailed records of the data concerning the strains received, and supplying these data on request.
5. before dispatch, fungus cultures should be carefully checked as to purity and trueness to species.
6. prompt answers to requests.

7. issuing a catalogue listing the available strains, giving short
and clear information pertinent to each strain, giving this catalogue a world wide distribution and keeping it up to date by regular
new issues.

All the requirements to which a well functioning culture collection
should conform, are easily stipulated, but surely it is no easy
matter to fulfil this task. If the authenticity of cultures of yeasts
and bacteria were checked before their incorporation into the collection, it would not be necessary to do this each time they are
ordered. In a larger fungus collection, however, where often cultures are requested that may not have been examined by microscope for several years, we deem a careful examination essential
before dispatch, both for the benefit of the receiver and for the
reputation of the collection. To prepare and regularly distribute a
catalogue is a matter of concern to most curators of collections.
Apart from the funds needed for publishing the catalogue, preparing a new edition is time consuming, hence expensive. The ever
changing nomenclature has to be brought more or less up to date.
Bacteriologists refer in their catalogues to the nomenclature used
in Bergey's Manual of determinative Bacteriology. For yeasts
reference is made to the monograph of Lodder and Kreger-van Rij.
Changes in the nomenclature of filamentous fungi, however, are
published in many different mycological periodicals, and not rarely
in journals of general botany or microbiology. Moreover it is often
not easy to decide whether or not to follow a nomenclatural change.
Too many and too frequent changes should be avoided. On the whole
the editor of a catalogue should be rather conservative, otherwise
the list will be difficult to handle for non-taxonomists. Cross-
references should be given from the antecedent names to those
employed.

In order to ascertain what information is desired concerning
strains listed, I sent a questionnaire to microbiologists of several
industries in Europe and the United States. From their answers I
gathered that on the whole the information given by the catalogue
of the American Type Culture Collection was considered satisfactory. For information to be provided to each strain, I would submit the following scheme:

Scientific name of organism
Accession number
    - synonym
    - donor's name, with his registration number.
    - authentic strain, or derived from type material.
    - year of accession in collection.
    - substratum from which the strain was isolated, country
      and year of isolation.
    - short indication of pathogenicity, biochemical properties

or application, as authenticated by donor.
- reference to pertinent literature or patent number.
- accession numbers of the strain in other collections.
This list of data should be printed in as condensed a form as possible. However, a mere enumeration of numbers and abbreviations
that have to be deciphered like a code should be avoided. Uniformity in signs and abbreviations used, and also in the sequence
in which the data are given, is highly advisable, especially for the
catalogues of service collections. Passage of a strain from one
collection to another, might be indicated by the code letters of the
collections and the accession numbers of the strain, linked by an
equal sign. Concerning pathogenicity and biochemical properties
of the strains, it should be made clear in the introduction of the
catalogue that these properties can never be guaranteed, because
the general collections have no facilities for testing them.
Culture collections as centres of active study; cooperation
between specialized and service collections

The specialized collections, as already stated, are built up
around centres of active research. The staff members of a service
collection, however, are often so fully engaged in the duties involved in supplying other microbiologists with the cultures needed,
that little time remains for actual research. These staff members
live amidst a tantalizing wealth of material to be studied, and
enticing problems open up almost every day, but: giving service
has precedence. Fortunately at times publications do appear,
monographs are published, which add to the usefulness of the collection, and are stimulating to the workers. Most collections
welcome guest workers. If they are advanced in a special branch
of research, or in a taxonomic group, their visits can be very
inspiring to the staff of the collection, by their showing new
methods. By giving to the species of his interest a thorough over-
haul, the collection will always benefit.

Cooperation between the service collections is self-evident,
and, to my experience, lively. I would recommend a more intense
interrelation between service and specialist collections. The
larger collections may help the specialized collections to strains
they need, and on the other hand will be grateful to receive fresh
isolates and strains of industrial value. The general collections
are often asked about the actual pathogenicity or biochemical properties of available strains. They are, however, not equipped to
check them. They have no experimental fields with host plants,
nor specialized laboratories to screen their organisms. If, from
time to time, the physiological properties of an important strain
could be checked by specialists, this would greatly add to the
proper service of the collection.

It not infrequently happens that an organism used in research

no longer shows the properties expected, simply because it has
been replaced by a contaminant, which remained unnoticed by the
investigator. In this case the taxonomist of a service collection
might be consulted and asked to verify the identity of the organism.
This leads to the last and most important point I want to submit
viz.: The need of an identification service. The larger collections
with their wide range of species are a very good base for giving
this service. The taxonomists have at their disposal an extensive
literature, authentic cultures which can be used for comparison,
and the experience in their particular field. The number of staff
members has to be equated to this task. One single identification
may take days or even weeks of tracing old literature and careful
comparison with known species. Several collections are already
giving this service. To the Commonwealth Mycological Institute in
Kew, fungi and phytopathogenic bacteria can be sent for examination, in the first place by workers in the Commonwealth. The
Laboratoire de Cryptogamie in Paris and the Centraalbureau voor
Schimmelcultures in Baarn, identify fungi on request. The National
Collection of Industrial Bacteria and the National Collection of
Marine Bacteria, in Aberdeen, offer a similar service for non-
pathogenic bacteria. The American Type Culture Collection states
in their catalogue that they are at present unable to provide an
identification service for unknown cultures. A director of a culture
collection in California wrote in answer to my question whether
identifications could be carried out at his institute: "Yes, but we
prefer not to do so". I can very well understand this reaction, for
to ascertain the true identity of an organism, is a responsible part
in the work of the investigator, who is moreover often pressed for
an answer without delay. I think it is of paramount importance that
more attention be given to the further extension of identification
services. With the rapid development of industrial microbiology,
an ever increasing number of biochemists have to work with micro-
organisms. Taxonomy is often considered by them a rather out of
date branch of science, hardly worth any attention. Their organism
has a number and that suffices .... until their isolate proves to be
of importance and has to be mentioned in a publication or even in
a patent application. Then it will need a name and the taxonomist
is asked to provide it, on the spur of the moment! Microbiologists
in all the various branches of research need reliable identifications
of their organisms. For want of this, utter confusion will set in.
Specialized culture collections cooperating with service collections
will be essential for the solid base of taxonomy.

References

1. Buchanan, R.E. Typification in Bacteriology. Intern: Bull.
        Bact. Nomen. Tax. 12: 17-26. 1962.
2. Herb. I.M.I. Handbook. Commonwealth Mycological Institute,
        Kew. 1960.

DISCUSSION I

by S. T. Cowan
Central Public Health Laboratory
London, England

I will start by correcting misconceptions about the National Collections in the United Kingdom which Dr. van Beverwijk - and perhaps others - hold. The organisation of collections within the British Commonwealth follows recommendation's made in 1947 by a specialist conference. A Commonwealth Permanent Committee co-ordinates the activities of the National Committees set up in each country to establish collections. Among the recommendations was one that collections should be specialised (this was a recognition that jacks-of-all-trades have no place in culture collections), and, within any discipline (e.g. bacteriology) further broken down by functional specialities. Dr. van Beverwijk implies that the U.K. National Collections are local specialised collections; I can assure her that this is not so; the National Collection of Type Cultures is a collection of some repute internationally and it has been sending cultures abroad since its foundation in 1920; each year it sends 1,500-2,000 cultures to laboratories in other countries. We regard ourselves as a general or public collection (= service collection of Kluyver) which sends cultures anywhere and whenever asked to do so. In these collections all cultures must be checked, and our breakdown by function allows this to be done by experienced workers.

What I call specialised collections are much smaller and more intimate. They are collections made by the experts and the enthusiasts; as an example I think of the vibrio collection made by Bruce White; the shigella strains collected by E. G. D. Murray. Collections such as these are often lost when the expert collector retires or dies. The problem is how to prevent this loss. The general (service) collections cannot accept these highly specialised collections and make the cultures available on the usual conditions. Each collection has its routine for checking the purity and identity of strains sent to it and maintained by it (our routine involves 30-40 tests) and it would be impracticable to check the hundreds of special strains so thoroughly. But would it be possible to set up a collection within a collection? What I have in mind is this. Cultures would be divided into three categories: A, true type cultures; B, tested, typical cultures (these would be examined and authenticated by experts or by the Collection's own staff); and C, cultures received from experts, freeze-dried without any check by the collection staff, and issued without any guarantee of purity or correct determination. These C cultures would not be catalogued and would only be distributed to workers who want large numbers of strains of a particular species or genus; we could regard them as repository (at rest) cultures in distinction to the ordinary strains deposited for general distribution. But even this plan might involve a great deal of work for the collection staff, and the number of C cultures accepted at one time might have to be limited.

Dr. van Beverwijk has spoken of the desirability of issuing catalogues at regular intervals; we have found that it is much better to let workers tell us what they want, and we try to give it to them. There are many reasons against publishing catalogues, one is that people often refer to out-of-date editions; we still have requests from users of our 1936 edition but the contents of the National Collection of Type Cultures have changed since 1936.

Identification Centres

Colindale is a reference centre for the Public Health Laboratory Service in England and Wales and thus acts as an identification centre for bacteria and viruses that are known to cause disease, these are easily identified; the difficult cultures are those that do not fit any previous description; most of these are non-pathogenic. In the field of medical bacteriology we could employ a large staff trying to identify organisms isolated from man or animals; much of this work would be unrewarding for the clinician cannot wait the months it might take to find the answer. The person who isolates a culture should make his own identification; generally he will be able to do this down to genus, at which level determinative manuals may be of help. He should then obtain from a collection a number of known strains with which he can compare his unknown organism.

So much for comments on Dr. van Beverwijk's paper; I will now tell you what I think is the chief function of a culture collection. I, too, will speak from experience and confine my remarks to collections of bacteria.

To me, teaching material is the most important and it is the general (service) collections which are in the best position to provide it. A good collection knows what it keeps and has discarded all contaminated cultures. It has also checked the determinations of all its cultures and either corrected the faulty ones or discarded the cultures. This is time-consuming and it took us ten years to do it in the National Collection of Type Cultures. We are fortunate in being able to preserve by drying the bacteria we keep in the N.C.T.C. and it is possible, therefore, to examine ampoules of our dried products to make sure that what we have dried is what we intended to dry. We all make mistakes and occasionally mis-number tubes or make an error in recording, but these errors can be found by thorough checking of dried cultures, and the errant batch discarded. In the same way, we can keep observation on the viability of our dried cultures by making counts. Material that has been checked in the ways indicated is much more suitable for teaching than cultures that have been maintained on culture media.

I have said nothing of type material for this forms such a small proportion of our strains that it does not present great problems except in selecting the type cultures themselves, and that is a problem I cannot discuss here.

DISCUSSION II

by Richard Donovick
Squibb Institute for Medical Research
New Brunswick, U.S.A.

Dr. van Beverwijk represents a very important and highly respected culture collection which has, I am confident, often been of great service to all of us. This author makes an elegant plea for more extensive support for identification services which could be carried out by such culture collection laboratories. No one can reasonably disagree with this appeal. Like this author, I prefer to speak from experience and, therefore, would like to limit my comments to one type of specialized collection which Dr. van Beverwijk refers to as "the collections of fermentation organisms".

Such collections have grown up rapidly over the past twenty years stimulated primarily by the search for new antibiotic-producing organisms. Secondary growth has developed as a consequence of developmental studies in efforts to increase productivity of specific biosynthetic products of particular interest to various industrial concerns. To understand industrial culture collections one should distinguish between the cultures obtained through primary isolation and those which have been developed for their productive capacity. The tendency towards "carefully guarding their production organisms and strains" to which Dr. van Beverwijk refers is more likely to be applied to these latter developed strains. While these strains are important from an industrial standpoint, they may actually be of lesser importance in studies in comparative taxonomy. Here, the primary isolates may be much more interesting and, in fact, much more readily available for study by others outside the specific industry.

It is surprising to learn that Dr. van Beverwijk's experience has been that "for exact identification of isolates the industries often have to appeal to taxonomists of the culture collections". Presumably, the author is referring to taxonomists of service culture collections, according to her own classification of collections. This certainly must vary from country-to-country dependent in part on the patent laws of that country. In the United States of

America, for example, upon applying for a patent on a given fermentation process or product involving a micro-organism, it is customary not only to deposit the culture in a widely recognized collection from which the culture may be readily obtained, but also to supply a taxonomic description of the culture. Under such circumstances, it would be surprising, indeed, to learn that the given industry would need to appeal to outside experts to carry out the culture identification studies. This certainly has never been the case in our own laboratories at the Squibb Institute for Medical Research.

It is true that descriptions of cultures involved in patent applications may not be published until the patent has been issued but culture collections accepting such cultures for deposit have available to them a source of information on such cultures which, I believe, they are not using or requesting. I am referring here to the taxonomic studies which at least the United States customarily are supplied at the time of the patent application. I see no reason why culture collections should not request this information from the depositor at the time the culture is accepted for deposit.

Once a patent has been issued, such production cultures are as readily available as any other culture in a service collection and, therefore, I must disagree with Dr. van Beverwijk's gloomy picture of industrial research laboratories carefully hoarding their cultures.

Another aspect of industrial stock culture collections which appears to have been overlooked is that not all organisms maintained in such a collection have to do with biosynthetic products and processes. A sound industrial collection will also contain many other types of organisms which may, in fact, be highly standardized. These may include carefully selected strains for use in in vitro bio-assay procedures, for in vivo chemotherapeutic studies, especially developed cultures showing either highly specific or widespread resistance to antibiotics and other chemotherapeutic agents, mutants showing very specific nutritional requirements, etc., etc. Such cultures are made readily available to all workers in microbiology and, indeed, our laboratories, as well as many others, I am sure, send out many such cultures free of charge. It is unfortunately true that such cultures are not listed in catalogues. This is so not because of any desire for secrecy but because of the laborious task involved in preparing catalogues; a task to which Dr. van Beverwijk has adequately referred.

I wish to return briefly to Dr. van Beverwijk's comment on taxonomic studies, or rather the lack of them, in industrial research laboratories. Examination of the facts will show that such a generalization is not only unsound but also unfair to some very fine taxonomic studies which have been carried out in such laboratories. One need only refer to such fine papers as those published by Tresner and Backus (1), Backus, Duggar and Campbell (2), Routien (3), or Goos et al. (4) to realize that such basic studies are being carried out in industrial research laboratories.

In closing, I would like to outline briefly the purpose and function of a good industrial stock culture collection, as I see them:

1.   To act as a depository for newly isolated cultures of potential biosynthetic interest.

2.   To collect such cultures whether as new isolates or from other stock collections.

3.   To carry out taxonomic studies on cultures of particular interest and to make comparisons with other apparently related organisms. (Thus, a stock culture collection will serve in patent applications in the U.S. and other countries where such identifications are required.)

4.   To conduct studies on improvement in methods of maintaining stock cultures.

5.   To act as a depository of specially developed or standardized cultures such as those used in bio-assays, antibiotic resistance or antibiotic "spectrum" studies, chemotherapeutic or nutritional studies, etc.

6.   Not only to supply cultures on request to their colleagues for their research but, also, to advise on related organisms which might be of interest.

7.   To cooperate with other stock culture collections in comparative taxonomic studies.

References

1. Tresner, H. D. and Backus, E. J. A broadened concept of the characteristics of Streptomyces hygroscopicus. Applied Micro. 4: 243-250. 1956.

2. Backus, E. J., Duggar, B. M. and Campbell, T. H. Variation in Streptomyces aureofaciens. Ann. N. Y. Acad. Sci. 60: 86-102. 1954.

3. Routien, J. B. A key to certain species of Streptomyces. Rev. Latinoamericana Microbiol. Suppl. 3: 23-51. 1959.

4. Goos, Roger D., Cox, Elsie A. and Stotzky, G. Botrypodiplodia theobromae and its association with Musa species. Mycologia 53(3): 262-277. 1961. (Central Research Labs.-United Fruit Co.)

GENERAL DISCUSSION

V.B.D. Skerman, Australia - I wish to draw attention to something which seems to have escaped us in culture collection work. Nothing seems to have been done about attempting to set up a centralized collection of descriptions of strains of micro-organisms which exist in culture collections. I refer here to the enormous collection of data that Dr. Kaufman has at Copenhagen - the tremendous number of handwritten descriptions of every strain of Salmonella deposited there - and this is not readily accessible.

On the matter of identification services - I feel that there is a case for the setting up somewhere of identification services which do not attempt to store cultures.

M. P. Starr, USA - The preoccupation with very clean cultures that has been noted by one or two speakers is, of course, shared by me. However, there is one intermediate step which I would urge upon the perhaps over-zealous cleaner of cultures. That is, before the cleaning begins, the culture with its derivatives should also be preserved as received. This has proved many times to be of inestimable value in equating what the describer had intended as the breadth of that organism.

This brings me to remark very briefly on Dr. Donovick's suggestion that those strains which have been highly treated for a particular purpose are of no interest to the general taxonomist. On the contrary, they establish for us the plasticity of the organism and thus are very important.

E. J. Beckhorn, USA - The apparent conflicts mentioned appear to be not really conflicts but point up the diversity of the functions of culture collections. Dr. Cowan's suggestion that one part of the collection be identified accurately, guaranteed and authenticated, would supply the needs of taxonomy. The third part - the functional part, no guarantee - would in many ways be the most useful. Particularly for those groups interested in the study of functions of large groups of organisms. Here taxonomy becomes important only when a particular strain is selected.

E. G. Simmons, USA - I would like to back to the hilt Dr. van Beverwijk's suggestions on identification services. I suspect that some have misunderstood her and are thinking that she believes that there should be a central organization that is saddled with this job. I suspect that she means that those of us who are specialists in any group should make ourselves known.

Miss van Beverwijk, Netherlands - My idea was that the staff of a culture collection never has as many specialists as is necessary for all of the identifications. So, if we could have the cooperation of specialists from other culture collections, we might send them the problem strains.

E. G. Simmons, USA - Dr. Donovick, in the depositing of patent cultures do you normally deposit, in service collections, the original culture or the one that you have manipulated and are so anxious to keep control of?

R. Donovick, USA - The patent law is pretty clear in the United
States. The culture must be capable of doing what you claim it
does - it doesn't have to do it well. Usually when there has been a
great deal of work done on a culture, the work is aimed toward
having the culture do the job well. That culture would be much
more difficult to obtain.

NOTE - The general discussion sessions were recorded on tape
and the comments reported herein are, as nearly as possible,
verbatim. Individual speakers were not given an opportunity to
edit their comments. Because of space limitations, many comments which did not appear to bear directly on the subject matter
under discussion have been deleted from the transcript. The onus
of responsibility for deciding what should or should not appear
rests entirely with the editor.

THE ORGANISATION OF A TYPE CULTURE COLLECTION

by J. M. Shewan
Torry Research Station
Aberdeen, Scotland

Let me at the outset give a brief historical account of how the
two National Type Culture Collections - The National Collection of
Marine Bacteria and the National Collection of Industrial Bacteria
- came to be located and maintained at the Torry Research Station,
Aberdeen. With this information you will be better able to evaluate
our views on the organisation and running of such Collections.

The National Collection of Marine Bacteria

When I joined the Department of Scientific and Industrial Research in 1935 there already existed a small collection of marine
bacteria, mainly from fish, instituted by my predecessor
Dr. Mary M. Stewart; and to these were added, up to 1939, several
halophiles mainly from solar salts and salted fish, the whole collection comprising about 200 strains. War conditions inevitably
took their toll and in 1945 it was found that all Miss Stewart's
strains and many of the halophiles were dead. Things might have
been left there; but our work on the microbiology of fish spoilage
forced us to set up a new collection, because we found it almost
impossible to relate our isolates either to those described in the
literature or to any in a determinative key such as Bergey's. In
other words, our collection grew out of the practical necessity of
having authentic cultures to hand which we could compare, under
our own conditions, with either fresh isolates or organisms isolated by other workers. Thus we were drawn inevitably into the
taxonomic field and the collection grew up under this stimulus.
Indeed it would be our opinion, that a type culture collection is best
maintained in a healthy state where the group responsible for it is
carrying out research preferably leading to a better taxonomy.

In 1949 the U.K. National Committee of the British Commonwealth Collection of Micro-organisms invited my Department, the
D.S.I.R., to assume responsibility for the maintenance of a Culture Collection of bacteria from fish and fishery products and the
position of our Collection was thus consolidated.

In 1956, the position had been reached where the taxonomic
work, so far confined mainly to bacteria from marine fish and
fishery products, could be usefully expanded to include marine
bacteria generally. Marine bacteria play an important role not
only in fish spoilage but also in the deterioration of submerged
materials e.g. nets, and ropes. They take part in the biological
cycles of elements in the sea and they may be useful indicators of
the biological status of water masses and of their movements in

the oceans. At this juncture the Development Commission provided
funds to place a Junior and a Senior Research Fellow of Aberdeen
University at the Torry Research Station. With this assistance it
became possible to broaden the basis of the Culture Collection
there so as to include bacteria from any marine source, as well as
to initiate some research work on the taxonomy of certain selected
groups of marine bacteria. Thus in 1957 the National Collection of
Marine Bacteria as such was instituted. Starting with a nucleus of
200 cultures, the Collection now comprises over 900, many having
been deposited within the past 2-3 years by workers from all over
the world. Last year, my Department took over full responsibility
for the Collection, appointing a Scientific Officer, who devotes almost one quarter of her time to the Collection and the remainder
to research work, and one Scientific Assistant, who is employed
full time on the Collection. In addition research staff working on
fish microbiology participate in the taxonomic research and in
checking incoming cultures to the extent of 2 man years.

    The organisation, at first, was relatively simple, mainly be-
cause we could regulate our intake of cultures, these being usually
organisms that we had ourselves requested from other workers on
the basis of their published descriptions. However, the Collection
is gradually becoming better known to workers in the marine field
and within the past two years the number of cultures offered annually by outside workers has been over 200. Fortunately, it has
been possible to reorganise services required in common by the
two Collections so as to be better able to cope with these numbers,
which stretch existing resources to the limit and intake has had to
be regulated.

The National Collection of Industrial Bacteria

    The other Collection, the National Collection of Industrial
Bacteria, was transferred to Torry Research Station in 1959 as a
result of internal reorganisation at the National Chemical Laboratory, Teddington, England. This Collection began as a result of a
recommendation of the U.K. National Committee of the British
Commonwealth Collections of Micro-organisms, which, as with our
Collection from fish and fishery products mentioned above, invited
my Department, the D.S.I.R., to assume responsibility for the
maintenance of a Collection of bacteria of industrial and general
scientific importance. It was established at The National Chemical
Laboratory in 1950, with a nucleus of some 200 bacterial cultures,
taken over from the National Collection of Type Cultures at the
Public Health Laboratories, Colindale, London, when it discarded
cultures of non-medical interest. During its ten years at Teddington the Collection was built up, until, on its transfer to Aberdeen,
it had become well established, enjoying a world wide reputation.
This Collection was much more widely inclusive in its types than

the National Collection of Marine Bacteria, comprising many
groups seldom encountered in our marine work. While the Marine
Collection was initiated and grew up under our care, the National
Collection of Industrial Bacteria joined it at Torry with its reputation, organisation and routine already established. The problems
of running under one roof two Collections, conceived and developed
against such apparently widely differing backgrounds, have presented a stimulating challenge.

Organisation of the National Collection of Industrial Bacteria and the National Collection of Marine Bacteria

How then do we organise and run these two Collections? It
appears to us that how this should be done must depend to a large
extent on the conception one has of the purposes of a Type Culture
Collection, even although these may never be completely fulfilled
in practice.

Purposes of a Type Culture Collection:-

Two obvious and important functions of a Type Culture Collection are to receive and despatch cultures.

(a) Receipt and Checking of Cultures:-

Some would maintain that it is the duty of a Collection to receive all cultures germane to its field and would reject any suggestion that a curator should have the power to refuse any culture sent
to his collection, which is the obvious repository for it. However,
it seems clear to us that the curator must regulate intake to avoid
being overwhelmed. This has happened with the National Collection
of Marine Bacteria; we have been offered at one time batches of
anything from 25 to 500 cultures!! Even if every culture had been
fully characterised and documented the burden of maintenance
alone would have been insupportable. However, in many cases, the
cultures offered are ill defined and have only been superficially
investigated. The depositor apparently wants someone to maintain
his cultures until he can find time to investigate them more fully.

We would maintain that there are only two certain grounds
for the immediate acceptance of a culture germane to a particular
collection. (1) When it is the subject of a patent and is required by
the law of the country to be deposited in a recognised collection;
and (2) when a valid description has been published in a reputable
scientific journal, the assumption being that the paper has been
subjected to proper scrutiny by some expert in the field. Some
would go further and propose that the organism itself, if it exists
in pure culture, should also be examined by a recognised expert.
There are sufficient instances in the literature within the past
decade of new specific names that would never have been proposed
had such a procedure been adopted. It should also be a condition
of publication that the organism, if it exists in pure culture, and

not subject to patent restriction, be offered to a recognized Culture Collection. This last requirement would appear to be indisputable; but it is surprising how often authors describing new species find it difficult to make their cultures available to other workers for comparative studies.

In a well organised Type Culture Collection every culture accepted, whether freeze dried or in an active state, should, immediately on arrival, be checked for viability and purity. To meet this requirement intake of cultures must be regulated. It is surprising how often both freeze dried and active cultures, cannot be resuscitated on arrival. In some instances this is due to ignorance of the properties of the organism, and as an example may be quoted our experience with a chromogenic bacterium isolated from sea water off the coast of Florida. On arrival, the active slope was placed, as was our usual practice with our own marine isolates, at 0°C in this instance over the weekend. On subculturing on Monday morning it was found to be dead. A fresh culture was immediately requested which on arrival was immediately plated out and found to be viable. These cultures, again placed at 0°C to await full examination, were, after 3-4 days, again found to be dead. On enquiry, the worker who isolated the organism, disclosed the fact that the organism was very easily killed at chill temperatures, something never previously encountered with our marine strains isolated from cold northern waters. This experience taught us that information must be obtained straight away from depositors, if not already sent with the culture, concerning the temperature of growth, media, and various other requirements of the organism.

Growth having been obtained, the next thing is to check purity. We find it difficult to accept the view of one early curator that it is not for the curator of a collection to question the purity or authenticity of a culture deposited by a reputable worker. It would be easy to maintain such a stand if there were no subsequent criticism of the purity of cultures from other workers later using them. It may well be that the mixed culture, and it only, had the properties claimed for it by the depositor; but unless this is pointed out when the culture is received it is safe to assume that the culture is believed to be pure. If, however, purity is not checked immediately it may be difficult later to apportionate blame between the depositor and the staff of the Collection. Without claiming to be faultless ourselves, we are bound to say that from recent experience not a few depositors seem to have been unable to overcome the difficulties, often very considerable, of isolating pure cultures.

Having obtained a viable, pure culture we next proceed to check its morphological, biological and biochemical properties as reported by the donator. Such data usually supply the information required for a decision to be taken regarding the taxonomic position of the organism; and it is also at this stage that any discrepancies between the depositor's data and our own become apparent and require to be ironed out. As all with experience in this field will know, these discrepancies may often concern items of fundamental taxonomic importance; some are undoubtedly due to differences in technique or interpretation or reading of the various tests. One need only recall the difficulties encountered by workers examining the strains of Streptomyces on behalf of the International Streptomyces Committee, when 90% agreement on the tests was obtained for only two cultures, to realise how difficult a matter this can be.

At this point the Culture is now given an accession number. A history card and data sheet are prepared containing all the data - see examples - and the culture got ready for long term preservation. Normally the first method attempted is freeze drying, using a freshly grown slope and "mistdesiccans" as the suspending fluid - sufficient to give 10-15 ampoules. Immediately after freeze drying the culture is checked for viability and purity. Ideally, a count by the Miles and Misra technique should also be done at this stage, because the percentage loss in viability can give some idea, albeit an imprecise one, of the expected storage life of the freeze dried preparation. It occasionally happens that freeze drying under these conditions fails; another technique e.g. storage under liquid paraffin, is then tried. Often the only solution is frequent transfer in active culture.

(b) Supply and Dispatch of Cultures:-

The second important function of a collection is to supply for teaching, research, industrial or other scientific purposes, cultures that are authentic in purity and type, and fully documented as to source, and whose properties and taxonomic position have been examined by a recognised expert and fully checked by the type culture personnel. At Torry, we try as far as possible to keep in close contact with known experts, pursuing research on the various groups held in our collections, to whom we can resort for information and advice. Such workers anyhow are almost certain to be using some of our strains and from time to time can feed back to the Collection valuable new data, as they accumulate.

A properly run supply service presupposes that staff concerned are thoroughly familiar with the organisms under their care; and maintain high standards of accuracy. With large Collections such as the National Collection of Industrial Bacteria it is obviously impossible for each worker to be familiar with every group. Hence we find it profitable to subdivide the collection among the assistants as shown in Table 1. This division takes account not only of sheer numbers but also of the frequency with which cultures

TABLE 1

Distribution of Effort in the National Collection of Industrial Bacteria

Research Assistant A.

Acetobacter)	100	strains
Acetomonas)		
Azotobacter	32	"
Escherichia	60	"
Aerobacter	26	"
Proteus	19	"
Spirillum	21	"
Others	20	"
	278	" Total

Research Assistant B.

Bacillus	151	strains
Clostridium	61	"
Desulfovibrio	25	"
Cellulomonas	12	"
Others	18	"
	267	" Total

Research Assistant C.

Pseudomonas	125	strains
Flavobacterium	20	"
Arthrobacter	17	"
Brevibacterium	10	"
Mycobacterium	12	"
Serratia	11	"
Vibrios	10	"
Others	21	"
	226	" Total

Research Assistant D.

Lactobacillus	142	strains
Leuconostoc	34	"
Pediococcus	10	"
Aeromonas	19	"
Achromobacter	11	"
"Bacterium"	11	"
Others	24	"
	251	" Total

Research Assistant E.

Streptomyces	68	strains
Streptococcus	60	"
Chromobacterium	48	"
Micrococcus	22	"
Staphylococcus	10	"
Propinobacterium	34	"
Sarcina	10	"
Others	17	"
	269	" Total

are in demand, so that the routine of preparing fresh batches for
freeze drying is borne as equitably as possible. Each research
assistant is thus required to be responsible for the checking, recording etc. of incoming cultures assigned to his group as described above; to be familiar with organisms already in his care; and
to have stocks of cultures ready to be dispatched on demand. On
our present experience we consider that the maximum number of
cultures that can be assigned to an assistant of average quality
should not exceed 350.

All research assistants, responsible for cultures should be
thoroughly familiar with the freeze drying technique, and able to
execute it correctly. However, it has been our experience that this
task is more efficiently and expeditiously carried out when dele-
gated to one assistant not responsible for a set of cultures.

Requests for cultures are received every day. Each of the
five research assistants deals with these requests in rotation for a
week at a time. The aim is for all cultures to be despatched on the
day the order is received except in certain cases where charges
have to be made. The question of charges is almost certain to be
raised at this Conference and hence need not be fully elaborated
here. Suffice it to say that there are good grounds for believing
that few people realise how costly it is to run a type culture collection adequately. In view of the high cost a nominal charge for all
cultures can be reasonably defended on the ground that it would at
least deter some people from making unreasonable demands on the
collection. Anyhow there seems to be no good argument for not
charging industry the full economic cost of a culture. When a culture is requested from outside the U.K. and has to be charged for,
it is our usual practice to acknowledge the receipt of the order and
send an invoice for the amount due. Only on receipt of payment is
the culture sent. Bad debts and trouble with auditors are thus
avoided. Charging, of course, requires bookkeeping services,
which at Torry Research Station form part of the normal office
administration. Every culture of an organism dispatched, whether
charged for or not is entered on its history card (see examples).
In this way, we can see at a glance when stocks of freeze dried
ampoules are running low and which cultures are most in demand
and hence when the redistribution of cultures among the assistants
may be necessary so that they carry equal burdens.

Supplementary Functions of the Type Culture Collection

In addition to the two main functions just delineated there are
two other complementary ones which are of some importance. The
first of these is to act as a clearing house of information on all
aspects of the cultures held there. This would include not only
available data on the morphology, biology, bio-chemistry and physiology, industrial uses etc. of the organisms but also the best

methods for their long term preservation, their shelf life in relation to time and temperature conditions of storage, methods for
resuscitation, effect of methods of preservation on properties of
the organism and so on. The research assistants are thus expected to keep abreast of the relevant literature and to record references and other data on the history cards concerning their own
particular organisms. It is our usual practice to try to obtain re-
prints of all papers about work in which our organisms have been
used, as well as about techniques for the running of collections,
methods for the long term preservation of micro-organisms, effect
of storage on their properties, etc.

The other complementary function of a Type Culture Collection is to help to identify strains sent in by research workers, industrial concerns and the like. This is already an established
practice in some culture collections e. g. in the Centraal bureau
voor Schimmel-cultures at Baarn in Holland; and is one that we
have been attempting to undertake in a modest way in both our
Collections. Determinative work of this nature, even at the genus
level, is of course fraught with difficulties. Many groups are ill
defined taxonomically as anyone using Skerman's key will have
discovered. Even so it is usually a salutary experience for most
of us to try to determine the genus of an unknown organism.

Finally there is another matter which vitally affects the
health of a collection and which I have referred to earlier. We
consider it essential that the collection and its senior assistants
should be linked up with a research project of some nature. There
are two fields eminently suited for this purpose viz. taxonomy and
the long term preservation of cultures. As already stated we know
far too little about the failure of some organisms to respond to
freeze drying and other treatments, the storage properties of our
organisms, and the effect of freeze drying on the biological and
biochemical characteristics of organisms. Much of this work
should be tackled systematically for all organisms in a collection
and assistants should have time for such research. This we
attempt to do at Aberdeen.

We set great store on prosecuting an active taxonomic programme - in our case on the Gram negative asporogenous rods
such as the Pseudomonas, Aeromonas, Vibrio, Achromobacter,
Flavobacterium, and allied species. Here it is essential to have a
team working on morphology, biology, and biochemistry and physiology of these groups, using the Collection for reference purposes.
The organisation of such work at Aberdeen is shown in Table 2.

The work described in this paper was carried out as part of
the programme of the Department of Scientific and Industrial
Research.

TABLE 2

Taxonomic Studies Team

Scientific Officers

(Direction and Administration)

Morphology and allied fields	General Biology	Serology and Antigenic Analysis	Biochemistry and Physiology
(e.g. Electron Microscopy)	Routine microbiology and biochemical tests	Scientific Officer (Part-time)	Scientific Officer (Part-time)
1 Senior Experimental Officer	5 Research Assistants		
	(2 Experimental Officers and 3 Assistant Experimental Officers)		

301			
CAT. No.	SPECIES	STRAIN	ALT CAT No

COLLECTION OF   MARINE BACTERIA

DEPOSITED BY	J.M. Shewan AS		DATE
ISOLATED BY	W. Yaphe		DATE 1.11.52.
SOURCE	Isolated from Thodymenia palmata (a seaweed) collected at Point Pleasant Park, Halifax, N.S., Canada.		
DESCRIBED BY			DATE
SYNONYMS			
INFORMATION	Agar liquefier		

REFERENCES   Yaphe, W.   The use of agarase from Pseudomonas atlantica in the identification of agar in marine algae (Rhodophyceae)
Canad. J. Microbiol. 1957, 3, 987-993.

△ REVISE-RO

8118	Lactobacillus leichmannii	313 Tittsler L4SMcCoy NCDO 302	ATCC 7830
CAT. No.	SPECIES	STRAIN MODIFIED	ALTERNATIVE CAT. No.

NATIONAL COLLECTION OF INDUSTRIAL BACTERIA   HISTORY   CHECKED

DEPOSITED BY ATCC (on request)	CONFIRMED BY Dr. E.I. Garvie, N.I.R.D. Reading
DATE 16.3.50	DATE 10.3.61 (Batch 1957)
ISOLATED BY	RECORDED BY
DATE	DATE

SOURCE
INFORMATION >   Received as a freeze dried culture from A.T.C.C.   Used in assay of Vitamin B$_{12}$
Gives very good results - Barton-Wright July 1957.   Grown on "Difco" Micro-Inoculum
Broth for freeze-drying and resuscitation.

SYNONYMS >

REFERENCES	REQUESTS			
Hoffmann, C.E., Stokstad, E.L.R., Franklin, A.L.,				
and Jukes, I.H.	DATE	ORDER NO.	DATE	ORDER NO.
"Response of Lactobacillus leichmannii 313 to the	9.4.62	5009		
Antipernicious Anaemia Factor"	12.4.62	5016		
J. biol. Chem. (1948), 176, 1465-1466	2.5.62	5041		
	3.5.62	5082		

DISCUSSION I

by W. A. Clark
American Type Culture Collection
Washington, U.S.A.

The American Type Culture Collection (ATCC) was established as a private, non-profit corporation for collecting authentic strains of micro-organisms of importance in teaching, research and industry and preserving them in a genetically stable condition.

The collection at present consists of about 8,000 strains of a wide variety of micro-organisms: bacteria, pleuropneumonia-like organisms, fungi including yeasts, viruses of animals, plants and bacteria, cell lines, and a few algae and protozoa.

The Collection is governed through a Board of Trustees by the scientific societies in the United States that are primarily interested in microbiology: the American Association of Immunologists; the American Institute for Biological Sciences; the American Mycological Society; the American Phytopathological Society, the American Society for Microbiology; the American Society for Pathologists and Bacteriologists, the American Zoological Society, the Genetics Society of America, and the National Academy of Sciences - National Research Council. Each society nominates one of its members to the ATCC Board of Trustees, which Board then elects from these nominees its members, usually for two consecutive three-year terms. The Board meets annually to review and plan the policies of the Collection.

The ATCC is managed by its Director, who is responsible to the Trustees. The culture collection is subdivided generally into a Collection of Bacteria, a Collection of Fungi, a Collection of Cell Lines, a Collection of Animal and Human Viruses (this is the Viral and Rickettsial Registry), a Collection of Plant Viruses, and a Collection of Bacteriophages. Each collection in the ATCC is under the charge of a Curator, who is a specialist in his particular field. Each Curator is responsible for the effective collecting, long-term preservation and distribution of the strains in his collection. He also conducts research on preservation methods, and in characterization or taxonomy of the organisms in his care.

In addition to the various collections of the ATCC, there is an Information Center, which is responsible for collecting, storage and retrieval and distribution of data pertaining to strains in the ATCC, as well as other collections, and for frequent publication of the various lists and catalogues. A Business Office receives and processes requests for strains and is in charge of cash receipts and expenditures. A Facilities Department provides services such as media preparation, glassware processing, laboratory maintenance, and shipping of cultures. These departments are managed by the Information Chief, the Business Manager and the Facilities Chief, respectively.

The Collection is operated at present with funds derived from distribution fees and government research grants.

The various collections in the ATCC are being organized, reorganized, or reworked, and there are still imperfections. However, I shall describe the general model on which the various collections will be based.

Each collection will have an advisory group of specialist-collaborators which will meet periodically to help review and revise if necessary the contents of the collection. The curator will be a member of his advisory group.

Seed stocks for each strain accepted into the Collection will be prepared by a specialist and lyophilized or frozen in his own laboratory, then submitted in lots of 25-50 ampoules to the ATCC. Alternatively the strain may be sent to the Collection in the active state, prepared and lyophilized at the Collection, and a specimen sent back to the donor for checking. Viability, purity and characterization tests will be performed at the ATCC.

The seed stocks of authentic materials will be preserved under optimum conditions in the dormant state. Such conditions will be determined by continuing research. Cultures for distribution generally will be prepared at the ATCC by expanding material from one ampoule of frozen or lyophilized seed stock and relyophilizing it. In this way original seed stock can be conserved for many years.

DISCUSSION II

by William C. Haynes
Northern Regional Research Laboratory
Peoria, U.S.A.

Dr. Shewan has presented an informative and engaging account of the historical background and organization of his collections. I find the history interesting, but I cannot embellish it, and so I shall proceed to the principal subject--

First, however, I shall describe briefly the structure of the ARS Culture Collection. It is administered by Dr. C. W. Hesseltine who, with Dr. John Ellis, has the care of about 3,000 cultures, primarily molds, some streptomycetes and a few higher fungi. Dr. L. J. Wickerham is in charge of some 3,000 yeasts, and I have the responsibility for about 4,000 bacteria, including a majority of the actinomycetes. The ARS Culture Collection is one of four investigation groups in the Fermentation Laboratory, which is one of the six laboratories comprising the Northern Division of the Agricultural Research Service.

In his opening remarks and again in closing his presentation, Dr. Shewan stressed the vital importance of research to the well-being of a culture collection. Speaking for our staff, I will add our voices in support of this thesis. The incentive offered by the opportunity to do research is perhaps the most important factor in attracting and holding adequately trained and enthusiastic personnel. Morover, the results of research, when published, contribute to the reputation of staff members and through them to the renown of their collection. The late Dr. A. J. Kluyver put it another way. He said, "The staff of a culture collection can never be restricted to mere technicians; without the active cooperation of fully qualified scientific workers, the collection will soon lose its usefulness"(1).

Dr. Shewan also pointed out that the kind of research is important. He mentioned two fields as eminently suitable--taxonomy and long-term preservation of cultures. I agree wholeheartedly because, again quoting Dr. Kluyver (1), "The bringing together of a great number of strains of very diverse origin offers an almost unique opportunity for a thorough study of a certain group of micro-organisms." The opportunity is equally unique for study of culture collection technology.

I can imagine that the value of research in these fields to the well-being of a culture collection will remain clearly evident to the administrators of independent culture collections. As Dr. Shewan's remarks make clear, it is possible also for the director of a dependent collection to keep his perspective in this respect. There is the danger, however, to a dependent collection that the objectives of the parent organization will so override those of the subordinate culture collection that the type or research allowed to its staff will have little or no relationship to taxonomy and culture collection technology. When that occurs, the staff members, even though they are "fully qualified scientific workers", will become increasingly deficient in knowledge of the cultures they are expected to maintain. I wonder whether anyone here feels that the type of research is irrelevant to the welfare of a culture collection?

Ahead of receipt and dispatch of cultures, which Dr. Shewan has cited as two main functions of a culture collection, Dr. Freeman Weiss (2) put conservation of cultures. I agree with Dr. Weiss, except that I would give equal preeminence to assembling and conserving cultures. By so stating the main purposes of a culture collection, I think the emphasis is transferred from the commercial aspects to the fundamentally more important scientific aspects. Furthermore "assembling" to me connotes an active acquisition of cultures rather than a passive acceptance of offerings.

To my mind the acquisition of cultures should include specialists collections even though an immediate demand for them is not anticipated. Again, Dr. Kluyver (1) wrote, "It will not need elucidation that for the progress of microbiology as a science it is of the utmost importance that micro-organisms which have been described and studied by one scientist will be at the disposal of other workers who wish to check or to extend the observations made." In keeping with this philosophy we have incorporated the collections of several prominent microbiologists into ours. Examples are the N. R. Smith collection of aerobic sporeforming bacteria, the Blakeslee mucors and the Taphrina collection of A. J. Mix. It might be thought that acceptance of large blocks of cultures such as these would place an unsupportable burden on the staff. In practice for us at least, no problem has arisen. It has always been possible to arrange either for periodic delivery of manageable batches of cultures or for delivery of cultures in a preserved state so that we could handle them at leisure.

By urging acceptance of cultures on a broad basis, however, I do not intend to disagree with Dr. Shewan's contention that a curator must have the prerogative to accept or reject cultures. He must; but I think the exercise of that prerogative will generally be in excluding cultures that are impure, that are of doubtful authenticity or those that fall outside the defined limits of the collection.

Under the operating conditions of his collections, Dr. Shewan has found a maximum of 350 cultures to be the number that an assistant of average ability can handle. We previously reported (3) about 1,000 cultures per person to be the proper assignment in maintaining a collection adequately. Differences in the operations of our respective collections are bound to explain the different estimates. I cannot be sure, but I suspect we rely to a greater extent upon lyophilization to preserve our cultures. With the exception of less than half a dozen strains, we preserve all bacteria and actinomycetes in lyophil. Better than 98 percent of Dr. Wickerham's yeasts are kept in lyophil, and approximately 90 percent of the mold cultures are also preserved in lyophil. Therefore, our staff is relieved to a large extent from routine transferring of stock cultures.

Another reason our people can handle a greater number and diversity of cultures is that we do not make a practice of reidentifying each accession, as I gather Dr. Shewan's people do. We examine cultures for purity and for general conformity to the salient characteristics of the family or genus named for the strain by the donor. Further than this examination we usually depend on the integrity of our depositors. In the course of 22 years we have seldom found our confidence misplaced. Let me hasten to say, however, that I would, if sufficient staff and facilities were available, prefer the safer methods advocated by Dr. Shewan.

All in all, it seems to me, Dr. Shewan operates his collections according to the same well-established principles that probably motivate all curators and directors.

References

1. Kluyver, A. J. On type culture collections. Report to the Bureau of the International Council of Scientific Unions. Delft, 1950.
2. Weiss, F. A. Maintenance and preservation of cultures. Chapter V in Manual of Microbiological Methods. Edited by the Com. on Bact. Technic., Soc. Am. Bact., McGraw-Hill Book Co., Inc., New York. 1957.
3. Haynes, W. C., Wickerham, L. J., and Hesseltine, C. W. Maintenance of cultures of industrially important microorganisms. Appl. Microbiol. 3(6): 361-368. 1955.

GENERAL DISCUSSION

R. Donovick, USA - With regard to the cultures accepted by various collections for patent purposes, do you accept cultures with no taxonomic descriptions whatsoever? I think that, if you do so, it is a bad mistake for the future of collections and for the future of taxonomy.

W. A. Clark, USA - I would like to pose a question to Dr. Donovick. What is the purpose of depositing this culture in a culture collection? Is it for taxonomic purposes or is it for use in some particular procedure? Obviously it is for use in some particular procedure. We try to exact from the various depositors as complete a description as they are willing to give us. I think that some move is afoot to demand a minimal description of the deposited strain.

R. Donovick, USA - I would not say that the only purpose for a collection is taxonomy. However, I think that there is a tremendous wealth of information that is being lost because, for patent purposes, taxonomic studies must be made. It is quite true that someone who wants a patent may not want to expand on a special function of the culture but that may be really a minor trait from the overall biological picture. I think that culture collections are losing this wealth of information since much of it does not appear in publications other than patents.

C. W. Christensen, USA - I would like to ask the curators if they have had any difficulty in freeing their stock cultures of phage. We have obtained cultures which have carried phages which have been somewhat latent and which have been very difficult to get rid of. When we have been able to free the cultures of phage they have behaved very differently.

W. A. Clark, USA - We have had great difficulty with some of our pseudomonas cultures.

FUNDAMENTAL ASPECTS OF CELL PRESERVATION

FUNDAMENTAL ASPECTS OF CELL PRESERVATION

INTRODUCTORY REMARKS

by T. O. Wikén

Laboratory of Microbiology, Technological University,
Delft, Netherlands

The microbiologists have made us familiar with an immense
number of different types of organisms capable of utilizing the
most diverse chemical compounds as substrates for the dissimila-
tory and assimilatory processes involved in building up and maintaining the living substance. These processes result in the formation of an enormous variety of extracellular metabolic products in
addition to the numerous compounds constituting the cell material.
As a matter of fact, the great value of the microbes to general
biology, industry, etc. is closely connected with the vast diversity
of chemical conversions which may be brought about by these
organisms. The application of microbes for industrial purposes is,
furthermore, based on the existence of several forms in whose
metabolism a certain conversion occupies a predominant position
which results in the formation and accumulation of a particular
end product of commercial value.

The number of industries based on the metabolic activities
of microbes is expanding year by year as new processes are developed and new microbial strains are isolated and tamed for the service of mankind. In view of this and other facts it may be said
without overstatement that the most valuable working capital of a
microbiological institution is its collection of stock cultures of
well defined microbial strains constant in their ability to produce
useful compounds in high yields.

During the last few years man has entered an era of extraterrestrial exploration, and it seems likely that there in the near
future will be manned space flights having a duration of months or
years, and perhaps even long-term extraterrestrial habitation. In
view of the fact that the mass as well as the volume of a space
cabin should be kept at a minimum, recycling or regenerating life-
support systems have to be constructed which simultaneously provide food and free oxygen and dispose of waste components including carbon dioxide. Such systems of biological nature should, in
addition to man representing the consumers, imply creatures of
three types, viz. producers like the microscopic algae, and decom-
posing and transforming organisms like the bacteria and fungi.
The significance of microbial cultures with well defined and

constant biochemical capacities for the reliability of these biological microcosms is obvious.

As is well known, the microbial strains maintained in our pure culture collections as stock cultures on various substrates may undergo spontaneous mutation. In consequence, a single-cell culture after several transfers in the course of months or years may be found to contain one or more new types of organisms in addition to the original isolate, and it even happened that this isolate after some months of subculturing, due to the process of selection following spontaneous mutation, was completely replaced by one or more mutants. In such cases the phenomenon of mutation mostly had unpleasant consequences in so far as the new microbial strains had lost the capacity of the original isolates to form one or more commercially valuable products in high yields.

On the other hand, the method of deliberately induced mutation by the aid of irradiation or mutagenic chemicals has become the routine means of obtaining microbial strains producing antibiotics and other complex organic compounds in enormously increased yields as compared to the parent strains, or showing nutritional deficiencies and therefore being of great value as test organisms in the elucidation of biosynthetic pathways and in the assays of amino acids, vitamins, etc. As a matter of fact, the analytical microbiology is a branch of science with ever increasing importance in research in general and applied biology, chemistry and biochemistry as well as in the routine testing and screening in industry. Of course, it is highly desirable to maintain the biochemical and physiological properties of the valuable mutants mentioned. The flexibility of microbial cells is a useful thing only when being under satisfactory control.

From the above it is evident that an extremely important but at the same time difficult duty of a curator of a culture collection is to keep the microbial strains true to type. In fact, a great number of empirical methods are available for the maintenance of stock cultures, e.g. the periodic transfer of growing and multiplying cells on solid media; the storage of cells under a layer of sterile mineral oil; the storage of spores or vegetative cells in sterile soil or sand; the deep-freeze storage, viz. preservation at temperatures varying from -18°C to approximately -190°C; and the lyophilization. These methods and their numerous modifications have one feature in common, viz. that they cannot be applied with the same success for preservation of all kinds of microbes. In view of the marked diversity in the morphological and physiological properties of various microbial species this seems quite natural.

In this session we will deal with the behaviour of constitutive and adaptive enzymes during short-time storage of adapted

and unadapted cells of a strain of <u>Saccharomyces</u> <u>pastorianus</u>;
with the mechanism of injury in microbial cells during freezing
and freeze-drying; and with the results obtained in applying
various methods in preservation of fungal cultures.

I am convinced that it will be a profitable and pleasant
session.

LOSS OF ADAPTIVE ENZYME DURING STORAGE

by S. G. Bradley
University of Minnesota
Minneapolis, U.S.A.

Introduction

Induced enzyme formation is dependent upon four distinct factors: the genetic potential of the cell, its physiological state, the presence of inducer in the milieu and the absence of inhibiting substances (1). The genetic potential for synthesis of an adaptive enzyme system is subject to the same selective and mutagenic actions of the treatments used in long-term culture preservation as are genes for constitutive enzymes. Maintenance of a given state of well-being is the objective of short-term storage, and is particularly important in biological assays where repeated sampling of the same indicator strain is desirable. The function of the inducer seems to be activation of pre-existing enzyme-forming systems by negating the effects of intracellular inhibitors (2). Retention of an induced enzyme is controlled by the same factors as the adaptive process itself (3).

Saccharomyces pastorianus possesses constitutive enzymic systems for fermentation of glucose and sucrose, and an adaptive system for fermentation of maltose (4). This report is concerned with the short-term storage of adapted and unadapted cells.

Results

Effect of freezing on constitutive enzymes. S. pastorianus grown in glucose medium (5) was harvested by centrifugation and suspended in the original glucose medium or in deionized water (5×10^8 cells/ml). The yeast suspensions were stored at 10° or -20° for 1-7 days; whenever indicated, samples were transferred to a 30° incubator. After the sample reached 25-30°, the yeasts were harvested by centrifugation, washed with deionized water, diluted to a population density of 10^8 cells/ml and their ability to ferment sugar was determined manometrically (6). Cells frozen in the culture medium were killed and lost enzymic activity whereas cells frozen in water or stored at 10° remained viable and metabolically competent for 1 week (Table I). After 15 days, glucose-grown cells suspended in water at 10° possessed two-thirds of their original glucozymase activity whereas yeasts in glucose medium were nearly as fit as fresh cells. Cells frozen at -20° for 15 days were inactive, irrespective of the composition of the storage milieu.

Retention of adaptive capability. Glucose-grown cells were harvested, washed, suspended in water, expended glucose medium or fresh maltose medium and placed in a 10° refrigerator or in a

TABLE I

Effect of short-term storage at 10° and -20° on activity of constitutive enzymes of Saccharomyces pastorianus

Temp. °C	Suspending milieu	Enzymic activity μL CO_2 evolved/hr		% survival
		glucozymase	invertase	
10	water	920	600	95
10	glucose medium*	980	650	95
-20	water	800	550	85
-20	glucose medium*	10	5	2

Enzymic activity was measured manometrically: 10 mg dry weight of glucose-grown cells, 20 mg carbohydrate and a nitrogen atmosphere. Results were essentially the same for cells stored 1-7 days.

*the original culture medium

-20° freezer. Subsequently, the samples were transferred to a 30° incubator and allowed to warm to 25-30°. The yeasts were sedimented by centrifugation, suspended in aqueous maltose + peptone and shaken at 30° for 2 hours. The cells were washed free of inducer and enzymic activity was measured manometrically. S. pastorianus frozen (-20°) in old glucose medium was inviable and unable to produce the maltozymase system, whereas cells frozen in fresh maltose medium or in water possessed substantial fermentative capacity. Yeasts stored at 10° in water, glucose medium or aqueous maltose were viable and able to adapt (Table II).

TABLE II

Effect of short-term storage at 10° and -20° on power of Saccharomyces pastorianus to form an adaptive enzyme

Temp. °C	Suspending milieu	Enzymic activity μL CO_2 evolved/hr			
		glucozymase		maltozymase	
		induced	uninduced	induced	uninduced
10	water*	980	950	660	5
10	glucose+	930	920	660	5
10	maltose‡	950	960	670	225
-20	water*	615	610	495	5
-20	glucose+	10	5	5	5
-20	maltose‡	825	825	545	5

Glucose-grown cells were incubated in 4% maltose + 1% peptone for 2 hours at 30°. Fermentative utilization of glucose and maltose was measured manometrically. Results were essentially the same for cells stored 2-7 days.

*also 1% aqueous maltose, glucose or sucrose
+original glucose culture medium
‡fresh maltose medium

Retention of adaptive enzyme. S. pastorianus was grown in maltose medium, harvested and suspended in water, glucose medium or maltose medium at 10° and -20°. After 1-10 days, the yeasts were warmed to 25-30°, harvested, washed and fermentative competence measured. Maltose-grown cells were killed when frozen in the original maltose medium; glucozymase and maltozymase activity was lost. In fresh glucose medium at -20°, viability and metabolic capability were maintained. Adapted cells stored at 10° retained full glucozymase and maltozymase activity but cells in aqueous maltose, maltose medium, aqueous glucose and glucose medium showed progressive deadaptation (Table III).

TABLE III
Retention of adaptive maltozymase activity

Temp. °C	Suspending medium	Enzymic activity μL CO_2 evolved/hr	
		glucozymase	maltozymase
10	water	880	575
10	2% aqueous glucose	880	180
10	2% aqueous maltose	870	380
10	glucose medium*	860	80
10	maltose medium‡ *	850	265
-20	water	745	420
-20	glucose medium*	810	515
-20	maltose medium‡	15	0

S. pastorianus was grown in maltose medium, harvested and suspended in the appropriate medium. Fermentative activity of 10 mg dry weight of cells was determined manometrically. Results were essentially the same for cells stored 3-7 days.

* fresh medium
‡ original medium

Factors affecting hardiness of yeasts stored at -20°. In previous experiments, glucose-grown cells were killed in glucose medium at -20° but not in maltose medium; conversely, maltose-grown cells died at -20° in maltose medium but not glucose broth. It was noted that glucose did not freeze until the temperature was reduced to $-8 \pm 1°$, maltose medium at $-2 \pm 0.5°$ and deionized water at $+0.2 \pm 0.1°$ and that freezing rates and thawing rates were affected accordingly. Therefore, glucose-grown cells and maltose-grown cells were harvested, washed and transferred to fresh glucose or maltose broth and the cultures stored in a freezer at -20°. After 1-7 days, the samples were withdrawn and thawed quickly (in a 30° water bath), slowly (in a 5° refrigerator), or at an intermediate rate (20° incubator). The rate of thawing had little effect on viability or enzymic activity. Adapted and unadapted cells retained full glucozymase activity; only adapted cells possessed maltozymase activity (Table IV).

TABLE IV

Effect of rate of thawing on retention of adaptive enzyme

Thawing time	Suspending medium	Enzymic activity, μL CO_2/hr	
		glucozymase	maltozymase
1 min.	glucose	915	650
	maltose	890	635
20 min.	glucose	900	640
	maltose	875	640
150 min.	glucose	855	590
	maltose	850	575

S. pastorianus was grown in maltose medium, harvested and stored at -20°. Fermentative activity of 10 mg dry weight of cells was determined manometrically.

Next, the effect of the rate of freezing on viability and enzymic activity was determined. Maltose-grown cells were frozen quickly (in a dry ice and acetone bath), slowly (-5° freezer) or at an intermediate rate (-20° freezer). All samples were stored at -20° for 2-5 days and were thawed in a 30° incubator. Rapidly frozen cells lost fermentative capability whereas slowly frozen cells remained fully active (Table V).

TABLE V

Effect of freezing rate on adaptive enzyme activity

Freezing time	Enzymic activity, μL CO_2/hr	
	glucozymase	maltozymase
0.5 min.	125	95
20 min.	885	605
90 min.	965	670

S. pastorianus was grown in maltose medium, harvested and stored at -20°. Fermentative activity was determined manometrically.

Age of the culture as a variable in resistance to cellular destruction as a result of freezing was examined. S. pastorianus cultures harvested 2 days and 1 day after inoculation were compared with a culture that had been diluted with an equal volume of fresh medium 4 hours prior to harvest. The power of young and old cells to ferment sugar was different (old cells were less active) but the effects of freezing were comparable; that is, cells in expended medium were killed as a result of storage at -20° whereas cells in fresh medium were viable and enzymically active.

A critical factor for satisfactory culture preservation was the nature of the suspending medium. S. pastorianus was harvested, suspended in the growth medium or fresh medium and frozen in a -20° freezer. Cells stored in expended medium died whereas those in fresh medium survived. Acidity was not the sole factor

responsible for the unsatisfactory results with old medium
(Table VI).

TABLE VI

Effect of storage milieu on retention of adaptive maltozymase at -20°

Milieu	pH	Enzymic activity	
		glucozymase	maltozymase
fresh	6.5	875	610
maltose broth	2.5*	855	605
expended	2.5	15	5
maltose broth	6.5*	215	160

S. pastorianus was grown in maltose broth, harvested, suspended
in the original culture medium or fresh maltose broth and stored
at -20° for 2-5 days. Fermentative activity was measured
manometrically.

* pH adjusted with 1 N NaOH or HCl

Selective or mutagenic action of freezing and drying. Yeasts
that had been frozen at -20° in water or culture medium were
plated onto complete growth medium and single colonies that developed thereon were picked. Sets of 20 colonies were analyzed;
each set had one of the following histories: glucose-grown yeasts
stored in fresh glucose medium; glucose-grown cells stored in
expended glucose medium; glucose-grown cells stored in water;
maltose-grown cells stored in fresh maltose medium; maltose-
grown cells stored in expended maltose medium; or maltose-grown
cells stored in water. The individual clones were tested for power
to produce adaptive maltozymase and constitutive glucozymase.
These were compared with glucose-grown or maltose-grown cultures that had not been subjected to -20°. All clones gave comparable results, that is, uninduced cultures possessed high glucozymase activity but no maltozymase activity whereas clones sub-
cultured in maltose broth expressed both enzyme-systems.

Selective or mutagenic action of freezing was not detected in
the small number of colonies analyzed, therefore a differential
medium was devised to detect mutants or selected cells lacking
power to form adaptive maltozymase. The medium is a modification of the well-known bacteriological EMB agar. The yeast-eosin-
methylene-blue-maltose medium is composed of 20 g maltose, 20 g
agar, 2 g KH_2PO_4, 1.5 g succinic acid, 1 g NH_4Cl, 1 g yeast extract,
0.25 g $MgSO_4 \cdot 7H_2O$, 0.1 g asparagine, 0.045 g methylene blue and
0.3 g eosin Y. Maltose positive colonies (presumably expressing
maltozymase) are black with a metallic sheen; negative colonies
(presumably lacking maltozymase) are pale blue or pink and do not
display a metallic sheen. Samples from glucose and maltose cultures were plated onto the EMB-maltose medium to give 100-200
colonies per plate; over 10^5 colonies of each type have been

examined and all were maltozymase positive. Similarly, no
maltozymase deficient varients have been found among 10^5 colonies derived from frozen cultures. In addition, glucose-grown and
maltose-grown yeasts were dried, stored at 10°, subsequently rehydrated and plated onto EMB-maltose agar. No maltose nonfer-
menting variants were found among the 5×10^4 colonies examined.

Discussion

Short-term storage is desirable whenever several populations need to be pooled to obtain sufficient material for a particular determination, whenever the manipulations involved in making
a specific preparation are tedious or prolonged, and whenever repeated sampling of a discrete population is indicated. Results of
biological assays, for example, are influenced by the physiological
state of the inoculum. The health of the seed-culture is influenced
by so many factors that exact replication is impossible. The observed metabolic activity of a culture is determined by the age of
its inoculum, number of input cells, composition of the medium,
incubation conditions, the assay procedure, and by minutiae such
as time of day and exposure to light.

Short-term storage can provide standard inoculative material and thereby increase precision in an assay procedure but only
in a few well-studied systems. There are no universally successful
methods; what is true for maltozymase in Saccharomyces pastorianus is not necessarily true for maltozymase in Candida stellatoidea or for susceptibility to the antibiotic cycloheximide.

The results presented here point out some of the significant
factors in short-term storage (as mentioned above, the particular
results are not generally applicable). The techniques employed
were intentionally simple, even crude compared to the sophisti-
cated techniques now available for long-term preservation (refer
to other reports given at this Conference).

Storage conditions were the most critical factors for maintenance of adaptive maltozymase activity or for inducibility. Cells
in expended culture fluid were more susceptible to freezing and
thawing than were cells in fresh medium, water or sugar solution.
It is obvious that 10° retards but does not stop metabolic activity
because unadapted cells in maltose medium developed significant
amounts of maltozymase, and induced yeasts in glucose medium
underwent measurable deadaptation. The requirement for an
exogenous nitrogen source for both adaptation and deadaptation
was shown by inability of glucose-grown cells in aqueous maltose
to produce substantial amounts of maltozymase, and by the stability of maltozymase activity of induced cells in aqueous glucose.
Best retention of induced enzyme was where both nitrogen and
energy sources were absent. This is consistent with the hypothesis
that repression of induced enzyme synthesis is the result of

competition for transport ribonucleic acid rather than by the presence of a specific ribonucleic acid repressor.

Good storage fixes the physiological activity of the input population; therefore the recovered cells can be no better than the starting material. Maximal power to adapt was found in cells in the very early exponential phase of growth whereas maximal maltozymase activity in S. pastorianus was in the mid-exponential phase of growth.

In this study, the mutation rate for loss of the enzyme was so low that selective and mutagenic actions of the storage conditions were not detected. Preservation treatments are most likely to bring about alterations of a population when it is initially heterogeneous, when the particular expression is controlled by many nuclear genes or determined by extranuclear genes - in contrast to a homogeneous population in which the particular expression is controlled by a single nuclear gene. Power to produce antimicrobial agents, for example, is notoriously labile and is indeed genetically determined by plasmids and polygenes.

Acknowledgment

This investigation was supported by a research grant E-4075 from the U.S. Public Health Service.

References

1. Cohn, M. and Horibata, K. Physiology of the inhibition by glucose of the induced synthesis of the β-galactoside-enzyme system of Escherichia coli. J. Bacteriol. 78: 624-635. 1959.

2. Hogness, D. Induced Enzyme Synthesis. In Biophysical Science --A Study Program. Edited by J.L. Oncley. John Wiley & Sons Inc., New York. 1959. pp. 256-268.

3. Robertson, J. and Halvorson, H. The components of maltozymase in yeast, and their behavior during deadaptation. J. Bacteriol. 73: 186-198. 1957.

4. Bradley, S.G. Relationship between sugar utilization and the action of cycloheximide on diverse fungi. Nature, in press.

5. Sussman, M. and Bradley, S.G. Mutant yeast strains resistant to arsenate and azide. J. Bacteriol. 66: 52-59. 1953.

6. Bradley, S.G. and Creevy, D. Induction and inhibition of ∝-glucosidase synthesis in Candida stellatoidea. J. Bacteriol. 81: 303-310. 1961.

DISCUSSION I

by Donald H. Braendle
Abbott Laboratories
North Chicago, U.S.A.

In this brief discussion I will confine my remarks to the freeze drying or lyophilization process of culture preservation.

The two major objectives of any preservation process are (1) the insurance of culture viability, and (2) the retention of its heritable characteristics. The prime importance of the first objective is obvious, since the second cannot be attained unless the cells are viable.

Recently there has been considerable attention given to methods of improving the viability of lyophilized cultures. It is apparent that there is no universal method of preservation suitable to all cultures. Even closely related cultures differ in their responses to the same preservation procedure. The genetic constitution of a culture, which is responsible for its physiological and metabolic behavior, is a factor that must be considered not only in respect to viability, but also in the stability of individual culture characteristics. Some culture characteristics are more stable than others. Some are so essential to a cell that their alteration or loss can be lethal. The loss or alteration of other traits not resulting in lethality, can still cause the microbiologist much concern. These characteristics often rival that of viability in the importance of their retention. Such characteristics for example, are those concerning antibiotic production, sensitivities to various toxic agents, special nutritional responses, or even virulence. Possession of one or more of these traits by a culture is frequently the reason its preservation is desirable. Maintenance of such heritable traits depends upon the successful control of mutation and adverse selection.

Dr. Bradley has already mentioned that the quality of a preserved culture can only be as good as the starting material. This point is especially significant with respect to the genetic purity of a cell population. The forces of selection can only act if there is more than one type of individual present. Careful testing, selection, and growth of cultures to obtain the highest degree of uniformity, prior to processing, can do much to reduce unwanted selection.

Whether selection or mutation occurs as the result of lyophilization and rehydration has not been clearly demonstrated.

Selection can occur whenever stress is placed upon a population. Stresses imposed upon a culture by freezing, drying, and rehydration expose the cells to selective forces.

A few apparent examples of variation in lyophilized cultures of Penicillium chrysogenum auxotrophs can be seen in Table I.

The lyophilization procedures used in each case were, as far as is possible, identical in every case. The data are typical of the responses obtained from these cultures, and appears to be repeatable. Though the total number of observations were limited, there appears to be an alteration in the composition of the cell population. The population shift though small (between 1-2 fold in the

TABLE I

Genetic Stability of Lyophilized Penicillium Auxotrophs

Culture	Nut.* Req.	Before Lyo		After Lyo		% Killed
		Cells Tested	Proto-trophs	Cells Tested	Proto-trophs	
1C2-3	cho$_1$	2.8×10^5	0	5.1×10^5	0	89
2C1-1	cho$_2$	3.8×10^6	0	4.9×10^6	1	35
2C1-2	arg	1.3×10^6	1	1.7×10^6	1	32
1C2-1	isl	4.7×10^6	1	9.8×10^5	1	70
2C1-3	nic	4.4×10^6	8	3.1×10^6	9	65
2C1-3	nic	2.3×10^7	9	1.4×10^7	9	94

*Nutritional requirements: cho = choline; arg = arginine;
isl = isoleucine; nic = nicotinamide

case of the nicotinamide deficient strain) does not appear related
to the amount of killing. Only a small shift should be expected,
however, otherwise this method of culture preservation would be
unsatisfactory. Experience has demonstrated otherwise.

Examples of results obtained from similar tests with lyophilized cultures of Nocardia lurida are given in Table II.

TABLE II

Genetic Stability of Lyophilized Nocardia Auxotrophs

Culture	Nut.* Req.	Before Lyo		After Lyo		% Killed
		Units Tested	Proto-trophs	Units Tested	Proto-trophs	
1L2-2	try	6.5×10^8	0	8.2×10^8	0	37
1L2-6	arg	7.0×10^8	0	8.8×10^8	0	32
1L4	arg	2.8×10^7	0	2.8×10^7	2	52
		3.3×10^8	0	1.6×10^8	6	76
		4.3×10^9**	0	2.6×10^9	0	25
		1.8×10^7	2	9.8×10^6	23	24
		4.0×10^7	11	2.2×10^7	23	32

** The last three cultures were grown for 4, 14, and 20 days prior
to processing.

*Nutritional Requirements: try = tryptophan; arg = arginine.

Only one mutant out of ten that were observed demonstrated
a population change following the preservation process. This mutant, 1L4 frequently, but not always, produced prototrophic colonies following lyophilization, even when their presence could not
be detected initially. This increase was particularly evident if the
culture was aged prior to processing.

No evidence was found for population changes occurring during the storage of lyophilized cultures for periods up to three
months, except for a slow loss of viability, which seemed to affect
the prototrophs as well as the auxotrophs.

The data presented represent only very limited observations, however, they are suggestive as to the occurrence of selection and mutation during culture preservation and reactivation. Because of the inconsistent responses and the low order of magnitude, a clear demonstration of selection or mutation is difficult. It may be, however, that the observed variability between microbial cultures with respect to viability can be extended to the stability of genetic traits in strains derived from a common parent.

DISCUSSION II

by W. E. Brown
The Squibb Institute for Medical Research
New Brunswick, U.S.A.

Despite the great gains that have been made in synthetic chemistry man has found that certain syntheses can be accomplished more efficiently and more economically by utilizing the biosynthetic abilities of micro-organisms. The industrially important products of biosynthesis range in value from the relatively inexpensive acids and alcohols derived from sugars, to the relatively expensive products typified by the antibiotics and steroids. The biosynthetic reactions vary in complexity from dehydrogenation of the steroid molecule to complete synthesis of the complex vitamin B_{12} molecule. The utility of fermentation processes to man has resulted in the establishment of large fermentation plants both within and without the pharmaceutical industry in most if not all industrialized nations of the world.

In the final analysis everything depends on the micro-organism and its biosynthetic abilities. The wild type of micro-organism in which the desirable biosynthetic potential is first recognized usually does not carry out the reaction to a degree of efficiency which justifies its commercial use. Consequently, highly specialized mutant strains must be developed at no inconsiderable cost in terms of time and labor. Although the superiority of a mutant is usually based on greater efficiency in conducting the reaction in question other metabolic and morphological characteristics which influence product isolation such as ease of filtration, elimination of pigment production, and elimination of closely related compounds may be equally important on the industrial scene.

Once a superior culture has been developed it is most essential that the culture be preserved for many months or even years, in such a fashion that no physiological changes occur. Further, the culture must be grown in such a fashion that its full biosynthetic powers are put to use. Neither variation nor population selection can be tolerated during culture preservation or during the multiple stages of inoculum propagation leading to final employment in a plant fermentor. Prevention of population changes can be a difficult problem particularly with those cultures which are genetically unstable or are heterogeneous in character.

Having established that variations in the characteristics of a culture lead to losses which cannot be tolerated by the industrial microbiologists, I would like to pursue the matter further by describing how we in the Squibb Division of Olin meet the problem. Olin operates fermentation plants in the United States, South America, Europe and Asia. In these plants a variety of products are manufactured including antibiotics, steroids and vitamins. In order to eliminate the original culture as a factor which contributes to variable performance, master inoculum is produced at a central laboratory in New Brunswick, N. J. and shipped to the various plants as required. To meet this need we have established a specialized culture collection, the members of which have been selected because of their superior ability to produce chemicals under industrial conditions. The collection contains from ten to fifteen different species representing molds, actinomycetes and bacteria with each culture being the culture of choice for a particular biosynthetic process.

Our philosophy of handling these cultures is based on the following four factors:

1. Rigid control of the procedures followed during culture preservation. The optimum conditions need to be determined for each new culture acquisition.
2. Rigid control of conditions of culture storage. These too may need to be determined for each new acquisition.
3. Critical evaluation of the culture at the beginning of the storage period and at frequent intervals during the storage period.
4. Design of fermentation processes to reduce selective tendencies to a minimum.

Every time a fermentation is set the initial inoculum comes from a stable source, usually a lyophil vial. The vial normally is one of several hundred sister vials in one lot; the lot is prepared by a standardized procedure and evaluated thoroughly before use. Thus, vials from the same vial source can be and are used for setting fermentations over a period of many months in plants situated on four continents.

The vial lot is prepared with elaborate precautions to ensure preservation of a representative culture, and is then evaluated before it is released for manufacturing use. Evaluation includes tests for freedom from contamination, for viability, and for productivity. The productivity tests are made at both the flask scale and at the 30-liter scale. Comparison with a previously accepted lot of the same culture forms an integral part of the evaluation.

Because of the variability of biological processes we have found
that it is not practical to prove in the pilot plant beyond all doubt,
that a new vial lot is equivalent to its predecessor. Thus, if a vial
lot demonstrates a productivity 90% or greater of that of the control lot at both the shaken flask and the 30-liter scales, it is submitted to the domestic manufacturing unit for final evaluation.
Here it is introduced into the production operation slowly. Once it
is established that no marked reduction in yield is associated with
its use, normally determined after 3 to 5 batches, the tempo of
evaluation is increased.

 Analysis of the plant data for acceptance of vial lots has
taken several forms and is primarily dependent upon the number
of production batches which are run. When one is concerned with
processes for which several hundred batches are run per year the
final decision is postponed until the results from as many as 30 or
40 batches are available. This assumes that the preliminary results obtained with the new vial lot are neither markedly superior
nor inferior to those obtained with the control vial lot. Obviously
if the production level involves under 100 batches a year, it is
necessary for practical reasons to make a decision on many fewer
batches.

 In meeting the needs of all plants we distribute approximately 2,000 lyophil vials a year. With most processes the "industrial
life" of the culture is only 1 to 2 years because of its replacement
by a higher producer. However there are processes in which the
same culture has been used for several years and I would like to
cite our experience with two of these to illustrate the success of
our operation.

 We have used vials from the same vial lot of Streptomyces
azureus for the production of the antibiotic thiostrepton over a
four-year period. In this period the vials have demonstrated only
a 50% reduction in viability during storage at 5°C. Most important,
performance in the manufacturing plant has not varied.

 The second example involves the use of Corynebacterium
simplex for the dehydrogenation step in the synthesis of the corti-
costeroid triamcinolone. Over a period of five years approximately
15 different vial lots of this culture have been prepared, one from
the other. Although the 1-dehydrogenase is inducible we do not use
a steroid inducer at any step in the preparation of the vials. Based
on all of the criteria we have applied the culture has not changed
during the period and current performance in the plant is identical
with that first observed with this culture in 1957.

 I have already inferred that the design of the process may
prevent or reduce to a minimum unfavorable results from a culture
which is either genetically unstable or is of a heterogeneous character. Process stability can be achieved in one or more of several

ways, e.g., by reducing the number of vegetative stages in inoculum production or by regulating the physical conditions or the nutrients so as to favor the more desirable components of the population. Continuous culturing operations obviously are only suited to those cultures which are homogeneous and are genetically stable.

It has been my objective in the foregoing to indicate the utility that certain micro-organisms have in our industrial economy and to indicate the type of measures which are taken to ensure that variation and selection do not impair that utility. As I have indicated the use of rigidly standardized procedures and continued surveillance enables man to realize the full biosynthetic potential of these industrially important micro-organisms.

GENERAL DISCUSSION

R. W. Barratt, USA - Dr. Braendle, I would think that the data on Penicillium could be interpreted purely as a chance phenomenon just as well as evidence for induction of mutation during the process of lyophilization because, as I observed the data, there is only one case out of ten in which you observed an increase in prototrophy as a result of lyophilization.

D. H. Braendle, USA - The data given on the slide were only from those cultures which repeatedly showed this type of reversion. The many cultures which showed no changes whatsoever were not included.

S. J. Webb, Canada - Dr. Bradley, was there an actual loss of ability to produce adaptive enzymes or was it just a delay phenomenon?

S. G. Bradley, USA - For the particular conditions described in the manuscript, it is an absolute loss. However, on aging under some conditions, you may see a delay in induction or a requirement for greater concentrations of inducer.

R. Vincente-Jordana, Spain - Can some of the observed effects be the result of the type of water used?

S. G. Bradley, USA - We have seen differences in behavior of different lots of water but not in this particular series of experiments. We use a deionizing procedure which seems much preferable to the usual distillation procedure. When we do bacteriophage analysis, water becomes a much more critical factor but that doesn't apply in the behavior of these yeasts.

N. Grossowicz, Israel - Dr. Bradley, in cases when you have simply increased requirements for induction, don't you have evidence for slight damage to the cytoplasmic membrane?

S. G. Bradley, USA - Obviously something has been damaged when we begin to see additional requirements. Whether this is an external or an internal membrane, I don't think that we are in any position currently to evaluate.

MECHANISMS OF INJURY IN FROZEN AND FROZEN-DRIED CELLS

by Peter Mazur
Biology Division, Oak Ridge National Laboratory
Oak Ridge, Tennessee

Introduction

Freezing and freeze-drying are logical techniques for main-taining viable cultures of cells and tissues and have been successful in many instances (4, 8, 33). There are, however, some cellular forms that have never been successfully preserved and many others in which only small fractions of the initial population remain viable after treatment.

For many purposes, a low percentage of survivors would suffice if it could be certain they were genetically identical to the original population. Variation and selection will be discussed by others at this conference, so let me merely point out that the surest way to minimize selection is to have a high percentage of cells survive. Moreover, if a preserved culture is to be used directly, rather than as a seed for new cultures, high viability may be a necessity. This discussion will assume, therefore, that the achievement of high survivals is desirable.

It would be convenient if I could give some general rules and recipes for achieving this goal; unfortunately, I cannot. About the only valid generalization is the following: If a population of cells can be made to survive cooling to below about -50 °C and immediate rewarming to normal temperatures, it can be successfully pre-served for any desired time; moreover, effecting the preservation is simple and relatively inexpensive, viz. store the samples in liquid nitrogen or air at -190 °C. This generalization, it should be emphasized, will not necessarily hold at temperatures above -100 °C, at which many physical reactions and some chemical ones can still occur (5, 22).

The problem then is to prevent death during cooling and during subsequent warming, and it is a problem that has been met in chiefly two ways.

The first and most common procedure is to find a suspending solution that yields high survivals. Although this approach has been primarily empirical, it has been rather successful in some instances. King of the protective additives is glycerol, and its powers need not be delineated to those attending this conference. But numerous other agents have been found to improve survivals, including a variety of sugars, amino acids, poly-alcohols, polymers including polyvinylpyrrolidone, the newcomer, dimethyl sulphoxide, and a host of complex elixirs. The latter include mammalian sera, Naylor and Smith's medium, mist dessicans, milk-containing

additives, and the supernatants, washings, and extracts from cells
(4, 8, 33). One difficulty has been the uncertainty in knowing whether
an additive that successfully preserves one cell species will similarly protect other species. Not many investigators discard conventional culture preservation techniques without putting such an
extrapolation to the test--and often the test leads to a different
optimum suspending medium.

A second procedure for achieving high survivals is to manip-
ulate the physical variables. In freezing, this has meant a study of
the cooling rate, and much less commonly, of the warming rate.
More often than not, higher survivals have resulted from slow
cooling than from rapid, and the figure of 1°C/minute is rather in
vogue (19, 33, 35). But sometimes cooling rate is without effect (26)
and in yet other instances, rapid cooling is the superior (10). The
influence of the rate of warming for some reason has received
much less study, although a number of investigations have shown it
to be a critical factor (15, 17, 21). Here again, no generalizations
are possible, but rapid warming does appear to yield the higher
survivals in the majority of instances.

There is a third approach. Simply put, it is to find the physical mechanisms operating in cells during exposure to subzero
temperatures, to find out which of these are associated with or
cause death, and finally to choose conditions that will prevent the
lethal mechanisms from operating or at least minimize their lethal
consequences. Although simple to state, this mechanistic approach
is not simple to conduct and probably can only be approximated at
best. Still, to me it seems a useful approach, and the only one that
can yield findings of general application. Therefore, I would like
to devote most of the remainder of this paper to illustrate one
admittedly oversimplified way of analyzing freezing injury in
mechanistic terms and relate this analysis to some hypotheses and
experimental data.

Physical events occurring during cooling

As the temperature of a suspension of cells in an aqueous
medium drops below 0°C, the external medium at first remains
unfrozen and thus supercooled, but eventually begins to freeze at a
temperature determined by such factors as the volume of the solution, the type and number of nucleating agents, and the type and
concentration of solutes. When the external medium is water or a
reasonably dilute solution, the probability is much higher for
crystallization to be initiated extracellularly than intracellularly.
Cellular cytoplasm apparently contains few effective nucleating
agents, as evidenced by Salt's findings that most intact insect
larvae supercool to -20°C or below (28). Furthermore, if crystallization once begins externally, the released heat of fusion immediately raises the temperature of the suspension toward its freezing

point, and the rise in temperature still further reduces the likelihood of intracellular ice formation. Since heat is withdrawn by the
cooling bath, the temperature of the suspension eventually falls,
and the external water progressively freezes.

   As shown in Fig. 1, the vapor pressure of external ice falls
with temperature. The water vapor pressure of the unfrozen cell
interior falls also, but less rapidly. Therefore, as the temperature
drops, there tends to be an increasing difference in vapor pressure
between inside and out. However, the very existence of the differential sets in motion thermodynamic forces to eliminate it, and the
greater the differential, the greater the driving force for reestab-
lishment of equilibrium. Assuming an intact cell membrane, one
of two processes can occur. Either the cell interior can freeze, or
water can diffuse through the membrane and freeze externally.
The net result in either case will be the withdrawal of liquid water
from the cytoplasm and its deposition as ice (interiorly in first
case, exteriorly in the second). The resulting concentration of
solutes in the internal medium will lower its water vapor pressure
toward that existing outside. Although the two alternatives are
thermodynamically equivalent, they may or may not be equivalent

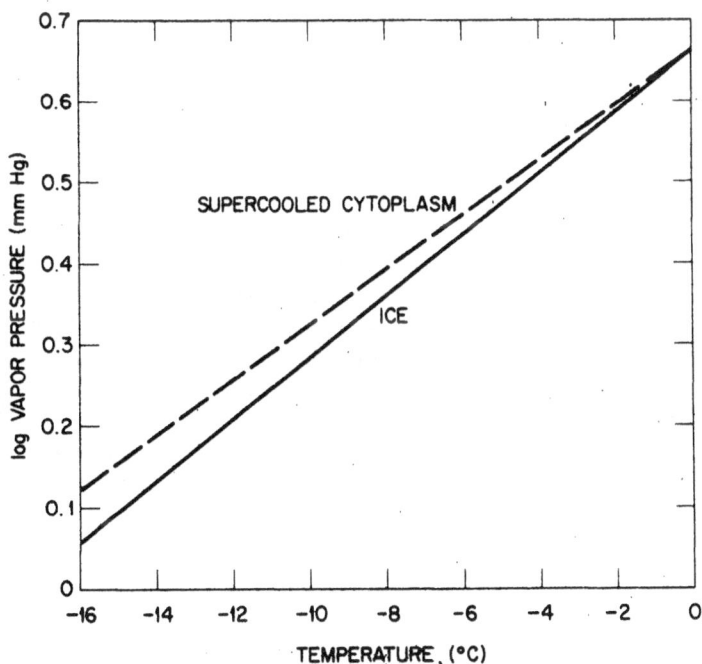

Fig. 1. Vapor pressure as a function of temperature.
The curve labeled "cytoplasm" assumes that cytoplasm
behaves like a 0.5 M supercooled, ideal solution.

in terms of injury. Both produce an increasing concentration of
internal solutes at a given temperature; and with a falling temperature, they produce an increase in both the internal and external
solute concentration. If this high solute concentration is a lethal
factor, either process would lead to death. On the other hand, one
process produces internal ice; the other does not, and conceivably,
the location of the ice could be more critical for survival or death
than the solute concentration.

Which will occur --intracellular freezing or loss of water
and deposition as extracellular ice? The answer depends on a complex interplay of factors, both biological and physical. Crudely
put, it will depend on whether it is easier or more likely for water
to diffuse outward than it is for the water to be converted to ice
in situ. More precisely, it will depend on the comparative activation energies for the two processes.

The rate at which water leaves the cell depends on the permeability of the cell to water, on the surface area of the cell, and
on the magnitude of the vapor pressure differential that is provid-
ing the driving force. Thus, for a given cell there will be an
inverse relation between total water withdrawn and the rate of
cooling. The slower the cooling, the more water will have left by
the time a certain temperature is reached.

Superimposed on these factors will be those determining the
likelihood of internal freezing. Firstly, the lower the temperature,
the greater the probability of ice being able to penetrate through
channels in the cell wall to initiate freezing (18). Secondly, the
less water that has left the cell by diffusion, the higher the freezing point of the protoplasm and the more it will be supercooled at
a given temperature; the more the cell contents are supercooled,
the greater the probability of their freezing (13).

The net result of these interrelations is that the higher the
cooling rate for a given cell, the greater the probability for intracellular freezing. And if various cells are cooled at a given rate,
the probability of intracellular freezing will be greater the lower
the permeability to water and the lower the surface-volume ratio.

The first of these deductions has been experimentally supported in several species of cells by a number of investigators;
that is, rapid cooling often yields evidence of internal ice crystals,
whereas slow cooling produces shrunken cells with no evidence of
internal ice (2, 23, 24).

Mechanisms of injury in frozen cells

It has been rather widely held that the basis of freezing
injury is the production of high solute concentrations by the
mechanism just discussed. The strongest support for this hypothesis came from an elegant study by Lovelock and co-workers
showing that the degree of hemolysis of red blood cells at various

subzero temperatures could be mimicked by placing red cells at
normal temperatures in sodium chloride solutions having the same
concentration as would exist at the given subzero temperature (10).[+]
Another way of expressing the hypothesis is that hemolysis occurs
at a critical electrolyte concentration, and therefore occurs at that
subzero temperature at which sufficient ice has formed to produce
the critical concentration in the residual unfrozen solution.

It followed from this hypothesis that any non-toxic solute added
to the suspension should result in less injury at a given subzero
temperature, for the presence of the added solute would reduce the
concentration of electrolytes at that temperature. A second requirement for a protective solute is that it be able to permeate the cells,
for only then would it produce the required dilution of the intracellular electrolytes. On the basis of this hypothesis, the excellent
protective effects of glycerol seemed understandable. It was non-
toxic to red cells and it could penetrate them (11, 12).

The hypothesis ran into two difficulties. On the one hand,
there was a tendency to assume that what applied to red cells, applied to cells in general. Secondly, the concept involved certain
logical consequences which are not supported experimentally in
many cells.

For example, the hypothesis leads to the prediction that rapid
cooling would be less harmful than slow cooling. This follows because injury from high solute concentrations would be primarily a
chemical process and, therefore, time dependent. Since rapid cooling would expose the cells to the concentrating solutes for a shorter
time it should be less harmful. But as I already have indicated, the
opposite is usually true; rapid cooling is more harmful. A comparison among three micro-organisms is shown in Table I (16, 20, 21).

TABLE I

Effect of cooling velocity on cell survival
(Warming at 1°C/min)

Organism	Minimum Temp., °C	Cooling Velocity °C/min	Survival %
Aspergillus flavus spores (16)	-75	270	5
		0.4*	38
Pasteurella tularensis (21)	-75	270	0.0006
		1	2
Saccharomyces cerevisiae (20)	-30	50	0.003
		1	35

*Cooling was at this rate to -50°C; then at about 20°C/min
to -75°C.

[+] Actually, hemolysis occurred only after the cells were suddenly
returned to isotonic saline.

Another concept, the requirement that a solute must be capable of penetrating the cell in order to be protective, does not hold in many instances. Taking yeast as an example, one finds protection conferred by solutes to which the yeast cell is highly impermeable. Some preliminary data are shown in Fig. 2. In addition, note that the solutes in these "protective" media are the very electrolytes which the "solute-concentration" hypothesis implicates as the lethal agents. Moreover, these media protect in spite of the fact that they withdraw intracellular water by osmosis and so produce a considerable rise in the concentration of intracellular solutes even before freezing has begun.

There are also a number of examples of protection by high molecular weight compounds that clearly are unable to penetrate the cell. In fact, such compounds as polyvinylpyrrolidone (27) and polyglycols (32) protect red blood cells, the very cells about which the hypothesis was made originally.

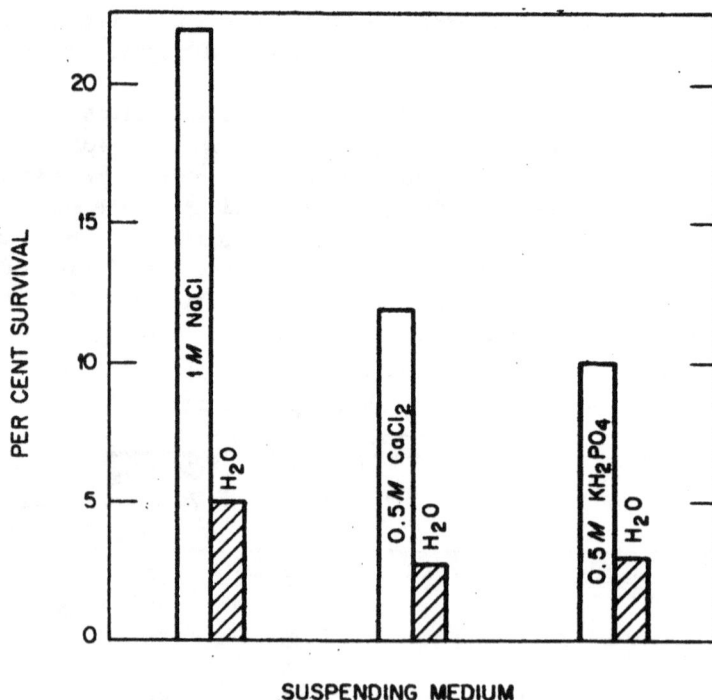

Fig. 2. Percentage of cells of S. cerevisiae surviving rapid cooling to -30°C and slow warming while suspended in the indicated media (Mazur and Holloway, unpublished data).

An alternative hypothesis of low temperature injury proposes that death is associated with intracellular ice formation. It too, leads to certain predictions; namely, that factors reducing the like-lihood of internal freezing should reduce injury. It was pointed out earlier that slow cooling should reduce the probability of ice for-mation within the cell; hence, the hypothesis predicts that slow cooling should reduce injury, and this it does in many instances (Table I).

Furthermore, if one were to partially dehydrate cells before cooling is begun, it would take them less time during cooling to reach a sufficiently low water content to prevent freezing than would be the time required for fully hydrated cells. Therefore, according to the hypothesis, the percentage of cells surviving a given rate of cooling should be higher for such partially dehydrated cells. Moreover, the percentage survival should increase with the extent of preliminary dehydration unless counterbalanced by increasing toxic effects from the dehydration itself. As shown in Fig. 3, this prediction is fulfilled for yeast cells which have undergone osmotic water loss by being suspended in solutions of sodium chloride, a non-penetrating solute.

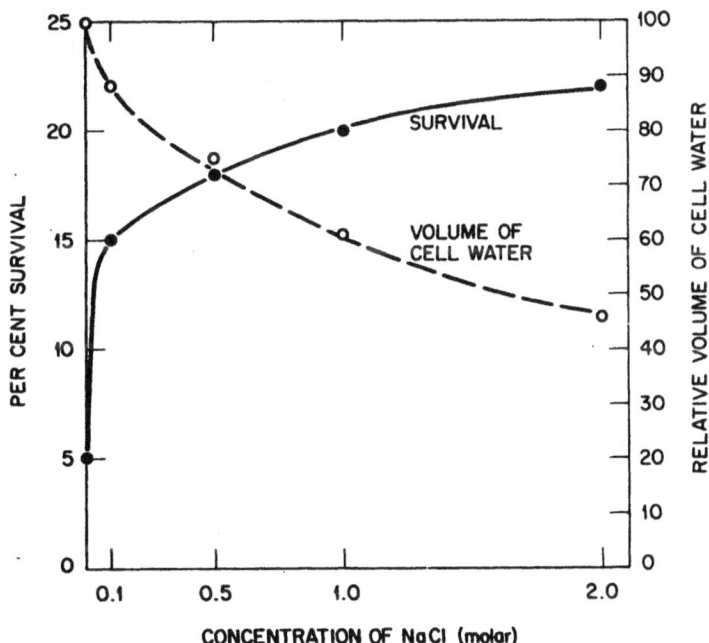

Fig. 3. Relation between the concentration of NaCl in the external medium and (a) survival after rapid cooling to -30°C and slow warming, and (b) the per-centage of water remaining in the cell before cooling (Mazur and Holloway, unpublished data).

Observations of a similar sort have been noted by Weston (36) and by Haskins (7). They found that a higher percentage of fungus spores survived freeze-drying if first subjected to preliminary dehydration from the liquid state.

I do not want to present and weigh here all the evidence on the concentrated-electrolyte theory of freezing injury vs. the intracellular-freezing hypothesis. Nor do I want to leave the impression that these are the only conceivable theories or factors; a number of others have been suggested, some of which are the following:

1. Thermal shock. This is injury resulting from a rapid fall in temperature independent of ice formation. In most instances, it has been neither demonstrated to occur or not to occur. Experimental evidence for thermal shock would require comparing the survival of cells supercooled to a given temperature with that of cells frozen at the same temperature and cooled at the same rate. These experiments are rare in the literature. One such comparison with yeast showed little or no injury to cells supercooled to temperatures at which freezing was decidedly harmful (20), and thus, eliminated thermal shock as a factor in injury in yeast. However, if a species of cells is highly sensitive to sudden chilling at above zero temperatures, the possibility would exist that similar injury at subzero temperatures could occur. One well-documented example of temperature-sensitive cells is bull spermatozoa (33).

2. Extracellular ice formation. The 10% increase in volume resulting from the conversion of water to ice might be thought likely to crush cells, but the evidence is against this supposition since many cells, especially micro-organisms, can in fact survive solidification of the external medium without harm (6,20,31).

3. Crystal size and form. It is well-known that crystal size decreases with increasing rate of cooling, and crystallization may even be incomplete at sufficiently high rates of the order of 1000°C/second (14,34). However, small crystals can grow during storage at temperatures above -100°C and during warming. Incompletely crystallized material can become completely crystallized [devitrification (14)] and larger crystals tend to grow at the expense of the smaller [recrystallization (22)]. In addition, the shape and habit of ice crystals can be affected by these same factors as well as by the nature of the solutes in a solution (14). These changes have been observed primarily in non-living systems, such as pure water and aqueous solutions. Presumably, they also occur in living cells, but how they affect survival is more a matter of speculation than of fact. About all that can be said is that factors which influence crystal size and form often influence survival in a way that would be expected. For example, rapid warming, which minimizes crystal growth by recrystallization or devitrification, also tends to yield higher survivals (18).

It is quite possible that several mechanisms contribute independently to injury in a given cell. Moreover, even if a single mechanism accounts for freezing injury in one species of cell, it by no means follows that it will explain injury in other cells, even those that are closely related taxonomically.

In spite of these complexities, there is still considerable value in thinking of the problem in mechanistic terms. Suppose, for example, one wishes to freeze a cell known to be highly sensitive to the intracellular and extracellular solute concentration or to partial dehydration. One might guess that the survival or death of such a cell would be determined primarily by the solute-concentration aspects of freezing; hence, cooling rates, suspending media and other variables could be evaluated in terms of the predictions from such a hypothesis.

On the other hand, suppose the aim is to preserve a cell that is rather resistant to changes in water content. The guess here might be that high survivals would be more likely attainable by selecting conditions to minimize the possibility of internal ice formation. Of course, the guesses may be wrong, either because the avoidance of one mechanism of injury may promote some other mechanism, or because the mechanism causing death is not what one guessed it to be. But even in these instances, one is no worse off than if conditions had been selected empirically. Actually, the failure of the cells to respond in a predicted way might well yield further information as to what mechanisms really were responsible for their injury.

Freezing vs. freeze-drying

My lack of reference to freeze-drying stems from the fact that it can be more ably discussed by Drs. Heckly and Nei. Before closing, however, I would like to point out some of the basic similarities between the two processes and also some of the differences.

Theoretically, once one has converted the cell water into ice, the process of dehydration has been effected, for at that point liquid water has been removed from the cell and sequestered in ice. The ensuing sublimation serves merely to remove the ice from the system, and should be less likely to cause injury than the initial dehydration produced by the freezing. If this were the full story, high survivals after freezing should yield high survivals after freeze-drying, and examples of such a correlation have been reported (1, 7).

Table II shows another such case (16).

However, in most instances studied, the correlation is not good; often a high percentage of cells remain viable after freezing, whereas few or none remain viable after lyophilization (3, 25, 30). Some of the possible sources of additional injury are the following:

TABLE II

Survival of A. flavus spores after freezing and
thawing and after freezing and drying (16)

Minimum Temperature °C	% Survival after		
	Freezing and thawing*	Freezing and drying	
		in H_2O	in horse serum
-15	65	25.7	45.2
-65	7	2.4	7.1

*Spores in water; cooled rapidly to indicated temperature and warmed slowly.

1. Final water content   In freeze-drying as ordinarily carried out with living cells, the cells are held at relatively high temperatures during sublimation and often are allowed to reach room temperature during the final stages of drying. Because of the great temperature differential between cells and vapor trap, freeze-dried cells should reach a considerably lower water content in a given period than cells in frozen suspensions at much lower temperatures, and these low water contents may be lethal.

2. Storage.   Freeze-dried cells are usually stored at above zero temperatures. Although such material contains insufficient water to support many chemical reactions, others apparently can occur at rates sufficiently high to produce a loss in viability with time (9, 29). On the other hand, when cells are stored at temperatures as low as -196°C, no biochemical reactions are likely to occur at measurable rates.

3. Rehydration vs. thawing.   Rehydration and thawing are analogous in that both make liquid water again available to the cells. However, the two processes differ in the rate at which water is made available and in the temperature at which they occur; thus, they may affect survival differently. Unfortunately, we do not know the mechanisms responsible for injury in either case, and, therefore, do not know whether similar or different mechanisms operate.

In spite of these differences between freeze-drying and freezing and thawing, the act of freezing remains a common factor to both processes, and no cell will survive subsequent thawing or dehydration if it has not survived the initial freezing. Accordingly, the more we understand the mechanisms operating on cells during ice formation, the greater should be the likelihood of devising procedures for successfully preserving the viability of a wide variety of cells and tissues.

References

1. Arima, J., Takahashi, Y., Nei, T. and Sato, T.   Japan. J. Tuberc. 1: 26-31. 1953.

2. Asahina, Eizo. Contributions from the Institute of Low Temperature Science, Hokkaido University, Sapporo, Japan, No. 10, pp. 83-126. 1956.

3. Benedict, R. G., Sharpe, E. S., Corman, J., Meyers, G. B., Baer, E. F., Hall, H. H. and Jackson, R. W. Appl. Microbiol. 9: 256-261. 1961.

4. Fennell, D. I. Botan. Rev. 26: 79-141. 1960.

5. Fernández-Morán, H. Ann. N. Y. Acad. Sci. 85: 689-713. 1960.

6. Harrison, A. P., Jr. and Cerroni, R. E. Proc. Soc. Exptl. Biol. Med. 91: 577-579. 1956.

7. Haskins, R. H. Can. J. Microbiol. 3: 477-485. 1957.

8. Heckly, R. J. In, Advances in applied microbiology. Vol. 3, Edited by W. W. Umbreit. Academic Press, N. Y. 1961.

9. Lea, C. H., Hannan, R. S. and Greaves, R. I. N. Biochem. J. 47: 626-629. 1950.

10. Lovelock, J. E. Biochim. et Biophys. Acta 10: 414-426. 1953.

11. Lovelock, J. E. Biochim. et Biophys. Acta 11: 28-36. 1953.

12. Lovelock, J. E. Biochem. J. 56: 265-270. 1954.

13. Lusena, C. V. Arch. Biochem. Biophys. 57: 277-284. 1955.

14. Luyet, B. J. Ann. N. Y. Acad. Sci. 85: 549-569. 1960.

15. Luyet, B. J. and Gehenio, P. M. Biodynamica 7: 213-223. 1955.

16. Mazur, P. Ph. D. Thesis. Harvard University, Cambridge, Mass. 1953.

17. Mazur, P. J. Gen. Physiol. 39: 869-888. 1956.

18. Mazur, P. Ann. N. Y. Acad. Sci. 85: 610-629. 1960.

19. Mazur, P. In, Recent research in freezing and drying. Edited by A. S. Parkes and A. U. Smith. Blackwell Scientific Publications, Oxford, England. 1960.

20. Mazur, P. Biophys. J. 1: 247-264. 1961.

21. Mazur, P., Rhian, M. A. and Mahlandt, B. G. Arch. Biochem. Biophys. 71: 31-51. 1957.

22. Meryman, H. T. and Kafig, E. Naval medical research institute project NM 000 018.01.09. August 16, 1955.

23. Meryman, H. T. and Platt, W. T. Naval medical research institute project NM 000 018.01.08. January 3, 1955.

24. Nei, Tokio. In, Recent research in freezing and drying. Edited by A. S. Parkes and A. U. Smith. Blackwell Scientific Publications, Oxford, England. 1960.

25. Nei, T., Hayashi, T., Sato, T., Ohara, Y., Nakagawa, I. and Maekawa, S. Low Tempe. Sci. B(Japan), 12: 63-70. 1954.

26. Rey, L. R. Proc. Roy. Soc. (London) B, 147: 460-466. 1957.

27. Rinfret, A. P. Ann. N. Y. Acad. Sci. 85: 576-594. 1960.

28. Salt, R. W. J. Insect. Physiol. 2: 178-188. 1958.

29. Scott, W. J. In, Recent research in freezing and freeze-drying. Edited by A. S. Parkes and A. U. Smith. Blackwell Scientific Publications, Oxford, England. 1960.

30. Sherman, J.K.　Am. J. Physiol. 190: 281-286. 1957.
31. Sherman, J.K.　Proc. Soc. Exptl. Biol. Med. 95: 543-545. 1957.
32. Sloviter, H.A.　Nature 193: 884-885. 1962.
33. Smith, A.U.　Biological effects of freezing and supercooling. Williams and Wilkins Co., Baltimore. 1961.
34. Stephenson, J.L.　J. Biophys. Biochem. Cytol. 2(Suppl.): 45-52. 1956.
35. Stulberg, C.S., Soule, H.D. and Berman, L.　Proc. Soc. Exptl. Biol. Med. 98: 428-431. 1958.
36. Weston, W.H.　Am. J. Botany 36: 816-817. 1949.

DISCUSSION I

by Robert J. Heckly and R.L. Dimmick
Naval Biological Laboratory, University of California, Berkeley
and
J.J. Windle
Western Regional Laboratories, Albany, California, U.S.A.

Many workers accept a survival of 1%, or less, for preserving stock cultures, but we should employ those methods that provide as near 100% survival as possible if all characteristics are to be maintained. The widespread use of lyophilization is evidence of its general applicability for preserving cultures, and Dr. Mazur's studies on the freezing of micro-organisms are of fundamental importance in reducing loss of viability on lyophilization. Initial recovery after drying, however, is only part of the problem of preserving cultures, because frequently the greatest lose of viability occurs during the storage period.

There are a number of factors that influence the survival of dried bacteria (3), and I do not intend to review these, but I would like to comment on some Dr. Mazur mentioned. Recent work on the effect of ultra high vacuum by Portner et al. (7) presents evidence that excessive drying, per se, is not detrimental. On the other hand, the manner of rehydration is frequently ignored as being an important factor, although it is well established that the activity of dry yeast is best restored by exposing cells to a humid atmosphere before use (8). Unfortunately, this procedure may not be generally applicable because Leach and Scott (5) showed that vapor rehumidification was highly lethal to Vibrio metschnikovii, and we found that under certain conditions vapor rehumidification also was harmful to Serratia marcescens.

As with frozen preparations, the temperature at which dried cultures are stored is an important factor in the survival of the cells. It is generally recognized that survival is inversely proportional to temperature, but there is no evidence that there is a critical temperature for storage of dried preparations as there is with frozen cultures.

Loss of viability is the change most often followed in bacterial preparations, but changes in other properties have been observed (3). As one example, we observed a reduction in virulence in Pasteurella pestis of more than 100-fold without a corresponding loss of viability (4). Apparently a large number of cells were injured during storage so that they could no longer grow and produce an infection in mice or guinea pigs, though most of the cells retained their ability to produce colonies in vitro. If these cultures were reconstituted and allowed to stand at room temperature, the original virulence was regained in about 24 hr, without any apparent growth.

Scott (9) has indicated that chemical actions occur in essentially dry preparations, and we believe that there must be some chemical or physical changes in the dry cells on storage if there is a loss of viability. In preliminary work with microrespirometers, we observed a slow gaseous uptake when freeze-dried S. marcescens was stored in air at various humidities; the rate of uptake increased logarithmically with increasing moisture content. The values, however, were not very precise, so we examined other techniques for obtaining measurements of changes in the dried state. We found that lyophilized S. marcescens produced free radicals as measured by an increase in the electron paramagnetic resonance (EPR) signal (1). Subsequently, Lion et al. (6) described free radical production by lyophilized Escherichia coli. As in our experiments, they demonstrated that the signal was produced only in the presence of oxygen, but they did not correlate signal strength with loss of viability.

We (2) found that the intensity of the EPR signal increased as viability of S. marcescens decreased. Fig. 1 shows a similar relationship between free radical production and viability of Sarcina lutea. The presence of lactose in the suspending fluid decreased both the free radical production and death rate.

Fig. 2 shows the effect of air on viability of Streptococcus lactis and free radical concentration. As with the aerobic organisms, a greater signal was produced in air than in vacuum, but the signal appeared to be smaller than that obtained with the other organisms studied. Free radical production may be limited by the fact that S. lactis has only a minimal capacity to utilize oxygen.

Dry yeast also produced free radicals in a manner that seemed to be correlated with death, and thus it appears that free radicals are produced by a variety of lyphilized micro-organisms. Apparently only the live cells are involved, because in no instance have we observed free radical production by cells killed with

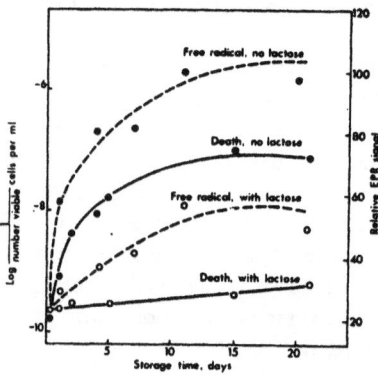

Fig. 1.   Effect of lactose on free radical formation and death of lyophilized Sarcina lutea stored in air.

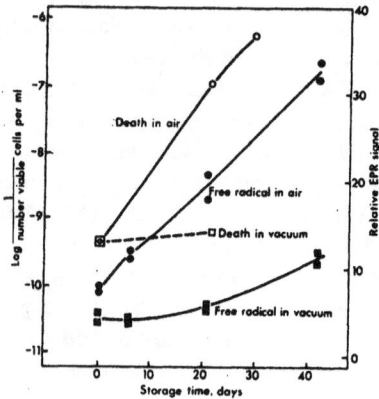

Fig. 2.   Effect of air on viability of free radical formation by Streptococcus lactis. Organisms were washed and suspended in 0.01 M phosphate buffer, pH 7.0, containing 0.5% lactose.

mercuric chloride, by heating, or by a poor lyophilization procedure. Therefore, we believe these signals to be a result of metabolic processes in the dry cells, particularly since it has been shown that free radicals are involved in certain enzyme systems. In the dry state free radicals are fairly stable and could accumulate if the mechanism for producing the free radicals is not blocked by drying.

Unless hyperfine structure is present, it is almost impossible to identify a free radical by its signal, but the fact that we have what may be a physical measure of death is intriguing. Of course, death and free radical accumulation could be coincidental, but we suggest that they are related by one or more of the following possibilities: a) the free radical is lethal during storage, or during rehydration, when the accumulated radicals are suddenly released to react with and overwhelm the cells; b) free radicals are a by-product of the death process; or c) dead cells react with oxygen to produce free radicals.

Although no free radicals accumulated in cells killed by mercuric chloride or heating, the results of a recent experiment, Fig. 3, suggest that free radicals were produced by cells which appeared to be nonviable. There was no change in the EPR signal of the cells stored in vacuum, although in the absence of lactose 90% of the cells appeared to have died. However, it is possible that these cells were still alive at this point and were just unable to grow on the media provided. When air was admitted to these "dead" cells, they produced free radicals at essentially the same

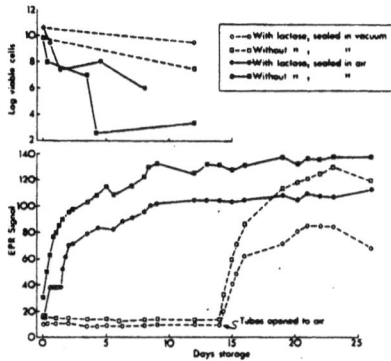

Fig. 3. Effect of air and lactose on
viability and free radical production by
lyophilized Serratia marcescens.

rate as did those that had been exposed to air immediately after
lyophilization. We believe that it is significant that the addition of
lactose reduced both the death rate and free radical production in
both instances.

Although the relationship between the free radicals and
viability is still obscure, we believe that EPR offers an interesting
and useful tool to enhance our knowledge of how and why lyophilized micro-organisms survive or die. By determining the mechanisms involved, comparable to Dr. Mazur's basic studies on the
mechanism of death by freezing, we will be able to minimize
losses in lyophilized preparations, and thus to develop a more
scientific methodology for preserving cultures than presently
exists.

References
1. Dimmick, R.L., Heckly, R.J. and Hollis, D.P. Free radical
   formation during storage of freeze-dried Serratia marcescens.
   Fourth International Conference on Medical Electronics, New
   York, July, 1961.
2. Dimmick, R.L., Heckly, R.J. and Hollis, D.P. Free radical
   formation during storage of freeze-dried Serratia marcescens.
   Nature 192: 776-777. 1961.
3. Heckly, R.J. Preservation of bacteria by lyophilization. In,
   Advances in Applied Microbiology, Vol. 3. Edited by W.W.
   Umbreit. Academic Press, Inc., New York. 1961.
4. Heckly, R.J., Anderson, W.W. and Rockenmacher, M. Lyophilization of Pasteurella pestis. Appl. Microbiol. 6: 255-261. 1958.
5. Leach, R.H. and Scott, W.J. The influence of rehydration on
   the viability of dried micro-organisms. J. Gen. Microbiol.
   21: 295-307. 1959.

6. Lion, M.B., Kirby-Smith, J.S. and Randolph, M.L. Electron-spin resonance signals from lyophilized bacterial cells exposed to oxygen. Nature 192: 34-36. 1961.
7. Portner, D.M., Spine, D.R., Hoffman, R.K. and Phillips, C.R. Effect of ultra high vacuum on viability of micro-organisms. Science 134: 2047. 1961.
8. Sant, R.K. and Peterson, W.H. Factors affecting loss of nitrogen and fermenting power of rehydrated active dry yeast. Food Technol. 12: 359-362. 1958.
9. Scott, W.J. In, Recent Research in Freezing and Freeze-Drying. Edited by A.S. Parkes and A.U. Smith. Blackwell Scientific Publications, Oxford, England. 1960.

DISCUSSION II

by Tokio Nei
Institute of Low Temperature Science
Hokkaido University, Sapporo, Japan

Dr. Mazur has described in his paper a number of experiments on the mechanism of cell injury resulting from freezing and drying, but many problems still remain. One important question to be investigated is the mechanism of freezing injury in cells of one species as related to another species. Micro-organisms have often been used as experimental materials in low temperature biology, but because of their size one encounters difficulties in investigating the morphological change in the cell during freezing and drying.

At this meeting time does not permit a detailed description of freezing and drying, which has been done by other investigators, so I shall only present some of the experimental data obtained at my laboratory in Sapporo. Yeast and E. coli cells suspended in distilled water were the experimental organisms used in this study. Morphological and physiological changes following freezing and drying and their effect on cell survival were investigated.

I. Freezing

1. Yeast Cells. A drop of yeast cell suspension on a cover glass was placed on a refrigerated microscope and the freezing process during slow or rapid cooling was observed (1). Cells cooled rapidly to -40°C at cooling rates of more than 10°C/sec retained their original size and shape, whereas cells cooled slowly at cooling rates of 1°C/min to the same temperature, became gradually shrunken as the temperature decreased. Those cells which retained their original size and shape i.e., the rapidly frozen cells, probably contained intracellular ice judging from their morphological appearance, while the shrunken organisms i.e., the

slowly frozen cells, were dehydrated during extracellular freezing. Cell viability following rapid cooling was markedly reduced when compared to those subjected to slow cooling. Therefore, it appears likely that the main cause of the cell injury due to freezing is the formation of intracellular ice crystals.

The results obtained from other experiments (2) on the effect of cooling rates on cell survival suggested that the percentage survival of the freeze-thawed cells depends primarily upon the cooling rate during freezing.

The fact that the percentage survival of the cells frozen at rates of about $10^3\,°C/min$ was less than that at 1 to 10°C/min could be explained on the basis of intracellular ice formation. On the other hand to explain the low survival at a cooling rate of $10^{-2}\,°C/min$, we must consider the effect of high intracellular solute concentration. We cannot explain the high survival at cooling rates of $10^4\,°C/min$.

2. E. coli Cells. Because coli cells are small in comparison with yeast cells, any internal morphological changes resulting during freezing must be observed with the electron microscope. In the specimens prepared by ordinary freeze-drying methods (3) or by freeze-drying in an electron microscope adapted with a special cooling device (4), rapidly frozen cells kept their original shape, while slowly frozen cells were irregular and distorted just as the yeast cells. Experiments in which the viable coli cells were determined after freeze-thawing, gave results similar to those obtained with yeast cells; that is, the percentage survival was less in rapidly frozen cells than in those subjected to slow freezing. Because of the difficulty in observing typical porous structures, such as seen in rapidly frozen yeast cells and interpreted as loci previously occupied by ice, we were unable to determine whether intracellular ice might play a predominant role in freezing injury of coli cells.

3. Comparison of Yeast and Coli Cells. As described above, morphological and functional changes in yeast and coli cells treated by freezing showed the same general trend, however, some differences were noted, as demonstrated by the fact that at the same high rate of cooling, most of the yeast cells contained ice and only a few survived freezing; whereas with the coli cells extracellular freezing occurred, and many of the cells survived freezing.

Such difference in resistance against freezing may be due to the different properties of the cells of the two species.

II. Water Content of Frozen or Freeze-Dried Cells

Freezable and unfreezable water content of yeast and coli cells as well as the residual water content of those cells after freeze-drying was measured (5). As shown in Table 1, most of the intracellular unfrozen water is withdrawn by freeze-drying. Yeast cells contain considerable water which can easily be removed by freeze-drying when compared to coli cells.

TABLE 1

Water content of frozen or freeze-dried cells

	Total water content			Dry matter
	Freezable water	Unfreezable water		
		Removed by freeze-drying	Residual water content after freeze-drying	
Yeast cells	69%	6.6%	0.4%	24%
Coli cells	60%	4.5%	1.5%	34%

III. Freeze-Drying

To investigate the mechanism of cell injury caused by freeze-drying, the relationship between the residual water content and the survival of the freeze-dried cells were examined in both yeast and coli cells (6). When the dehydration proceeded and the residual water content decreased below 7% in yeast cells and 6% in coli cells, which corresponds to the unfreezable water content of both cells, the percentage survival was rapidly reduced. Consequently, it was assumed that the unfreezable bound water might play an important role in cell viability.

References

1. Nei, T. Freezing process of yeast cells. J. Agr. Chem. Soc. Japan, 28: 91-94. 1954.
2. Araki, T. and Nei, T. Mechanism of freezing of micro-organisms. Survival of yeast cells subjected to subzero temperatures. Low Temp. Sci., Ser. B, 20: 1962 (in press).
3. Nei, T. Mechanism of freezing of micro-organisms. Morphological observations on Coli cells. (unpublished).
4. Nei, T. Electron microscopic study of micro-organisms subjected to freezing and drying. Low Temp. Sci., Ser. B, 19: 79-93. 1961.
5. Souzu, H., Nei, T. and Bito, M. Water of micro-organisms and its freezing. Low Temp. Sci., Ser. B, 19: 49-57. 1961.
6. Nei, T., Souzu, H., Hanafusa, N. and Araki, T. Mechanism of drying during freeze-drying. VIII. Relation between water content and survival of cells at different parts of material during the drying process (2). Low Temp. Sci., Ser. B, 19: 59-72. 1961.

GENERAL DISCUSSION

R. Donovick, USA - Dr. Mazur, are you confident that, during cooling, the loss of water is purely a matter of thermodynamics, or do you think that there is an active selective action?

P. Mazur, USA - No, I'm not really confident but I think that it is a thermodynamic state. Thermodynamics says that the water should go out and the morphological data show that the water does go out during slow cooling. No one has calculated how much water should leave and whether the rate at which it leaves is consistent with the morphological data. However, is there any data to suggest that water loss and uptake in micro-organisms is an active process and not just passive diffusion or an osmotic process?

R. Donovick, USA - My own thought is that we cannot assume that the solute concentration is going to be as high as you might think.

P. Mazur, USA - I do have some evidence that the membrane remains intact and solutes are not lost unless the cell is killed by freezing. If it is killed, the permeability barrier is disrupted. If it is not killed, the barrier remains intact and nothing leaks out.

A.P. Harrison, USA - Cells do not have to be dead to lose solute. In E. coli, for example, cells can lose 2% of their dry weight with no loss of viability.

P. Mazur, USA - I meant in yeast. I did not mean to imply that if any cell lost 20% of its solute, it was necessarily a dead cell. In our case, it seems to be pretty well an all or none phenomenon. I should add that one must be very careful to restrict any comments to a specific cell. It is dangerous to extrapolate to even closely related ones.

     With E. coli, I think the evidence is excellent that it is concentrated solutes that cause death. In yeasts, I believe that it is internal ice formation which is responsible.

B. J. Bloomfield, USA - Dr. Heckly, have you thought about the question of autooxidation of lipids as a possible cause of death of these micro-organisms?

R. J. Heckly, USA - I agree that this looks like it might involve lipids and work is progressing at the moment to test whether or not typical antioxidants will be effective.

P.A. Hansen, USA - There are two different processes generally used in lyophilization. In one the vials are melted off on the manifold. In the other, the tubes are first frozen and dried in a special container and later sealed inside a secondary container. From what I understand here, this second practice would probably be somewhat objectionable.

R. J. Heckly, USA - I would agree that the data which we have
would indicate that it is objectionable to allow air to reach the
dried bacteria for any length of time, particularly if the cells are
sensitive. However, I think that the convenience of the two-step
process probably outweighs the theoretical advantage of a single-
step process.

N. Grossowicz, Israel - Lyophilization is obviously a multiple
step process - freezing, drying and the rehydrating. I would like
to know of what importance is the last step on viability.

R. J. Heckly, USA - Rehydration is an important step, but it does
not overshadow the others. That is, if you have poor freezing, it
doesn't make any difference how you rehydrate. Conversely, you
can kill by methods of rehydration. This is one aspect that has
generally been more or less ignored.

E. G. D. Murray, Canada - I found that after freeze-drying delicate
organisms, such as neisseria, I could get a much greater recovery
rate if I put the cells into fresh blood first and then into ordinary
culture medium.

P. W. Muggleton, England - In the case of B. C. G., if the organisms
were freeze-dried under certain circumstances, a much higher
percentage viability is observed if the bacterial counts on recovery from freeze-drying are done on a medium containing 10%
blood. The difference, between with blood and without, is increased
the longer the cultures have been stored. It would appear that in
the freeze-dried cultures there is some essential substance which
is depleted by the organisms which blood will replace.

ASSESSMENT OF METHODS OF PRESERVATION

GENERAL METHODS FOR PRESERVING CULTURES

by Kenneth B. Raper
University of Wisconsin
Madison, U.S.A.

If I interpret correctly my role in this Conference it is to introduce in unspecific terms techniques that have been developed over the years for maintaining cultures of micro-organisms of special interest and significance. However, before considering some of the many methods that have been employed, and in part in lieu of such considerations, I should like to make some general observations and pay tribute to a few far-sighted microbiologists to whose vision and achievements this Conference might be dedicated.

The collection of biological materials and their preservation for purposes of reference must surely date back to Linnaeus, or even earlier. And the establishment and enrichment of herbaria on the one hand, and of comparable collections of insects, birds, mammalian pelts, etc. on the other, have represented important facets of botanical and zoological research for more than two centuries. True, such collections represented in the main desiccated and hence quite dead specimens useful only for comparative studies of pattern, structure, and morphology, including such detail as post-mortem microscopy could reveal. For obvious reasons, the collecting and cataloguing of lower organisms developed somewhat less rapidly, but by the early 1800's students of the fungi and the algae collected and dried their specimens no less enthusiastically than their phanerogamist colleagues. It was not however, until the latter half of the nineteenth century that such pioneers as DeBary, the mycologist, Koch, the bacteriologist and Hansen, the zymologist, developed techniques for propagating micro-organisms as pure and reproducible cultures in the laboratory.

The new and exciting frontiers thus opened led directly to the first collections of living micro-organisms, albeit these were limited in scope and represented only the special interests of the investigators. This pattern of isolation, study and maintenance of particular groups of micro-organisms by specialists during their active careers is still commonplace, and I would not discount this activity for even today such small collections represent a primary source of authentic and significant cultures. However, as emphasis shifted gradually from the characterization and systematization of related genera and species toward a deeper understanding of their roles as antagonists in disease, as despoilers of food, and as transforming and synthesizing agents in soil, the value of more inclusive and more permanent collections of micro-organisms

became apparent. More recently, their ever widening use as ob-
jects for basic research in physiology and genetics on the one hand
and as biosynthetic tools for industrial production on the other has
further emphasized their inestimable worth. It is largely in response to the evident and growing need for such collections that we
are now assembled. We have at long last come to realize that
micro-organisms, no less than higher plants and animals, represent an invaluable natural resource. But realize it we have, and
we meet here in a most favorable climate to exchange information
on how this resource can best be utilized and conserved, for the
microbes have attained status not only within the scientific community but in the eyes of the lay public as well. It was not always
so; some of our foremost collections of today had humble beginnings indeed! It is perhaps worthwhile to recount some of these.

The U.S. Department of Agriculture's internationally recognized NRRL Collection in Peoria, Illinois, had its origin in a laboratory at Storrs, Connecticut, more than half a century ago. A
young mycologist, Charles Thom, had undertaken the task of producing certain European-type cheeses in the United States. Being
a scholar as well as a cheese-maker, he was appalled at the confusion and misinformation encountered when he attempted to char-
acterize and identify the Penicillia he isolated from Camembert
and Roquefort cheeses. He read and collected avidly, and with the
warm encouragement of Professors Farlow and Thaxter at Harvard
University, where he journeyed to visit a mycological library, he
determined to "clean up the mess in Penicillium and Aspergillus".
It proved to be a herculean task, and when he moved from the Dairy
Bureau to the Food and Drug Administration and later to the Division of Soil Microbiology, he took his constantly growing collection
of molds with him. He early adopted the use of defined culture
media to secure reproducibility among the molds he assiduously
cultivated and compared, and with the aid of J. N. Currie he developed the first industrially practicable mold fermentation, i.e.
citric acid, in 1916-17. Subsequent to this he supplied the cultures
used by May, Moyer and others in their further investigations on
mold fermentations at the Arlington Farm Laboratory. The significant part of this story is that never in the thirty-eight years of his
employment by the Department of Agriculture was he authorized
officially to collect and maintain mold cultures, and the monographs
on Aspergillus (29) and Penicillium (28) were produced over and
above his assigned duties in the responsible positions he held.
With the opening of the four Regional Research Laboratories in
1940, the importance of a collection of agriculturally and industrially significant micro-organisms was at last recognized, and the
NRRL Collection was set up in the laboratory at Peoria. Two years
before Dr. Thom's official retirement his extensive collection of

fungus cultures, so long maintained as a labor of love, became the nucleus around which that large and diverse collection has been built.

The history of the Centraalbureau voor Schimmelcultures is equally interesting and no less significant for this Conference. The establishment of "a collection of living fungi for scientific research" was first discussed in 1903 at a meeting of the Association Internationale des Botanistes in Leiden and was actually established three years later at the Association's meeting in Paris. The moving spirit was Professor F.A.F.C. Went who contributed his own collection of fungi, numbering 80 cultures, as the nucleus of the enterprise. In the following year, 1907, the infant collection came under the dynamic and imaginative leadership of Miss Johanna Westerdijk, newly appointed director of the Phytopathologisch Laboratorium "Willie Commelin Scholten" in Amsterdam, and for the next half-century she served as its director. The Phytopathological Laboratory and the Centraalbureau were moved to Baarn in 1920, and rarely in the annals of science has the name of an individual been more closely linked with an institution than that of Professor Westerdijk to the C.B.S. It is quite impossible to estimate the contributions of this collection to the progress of science, particularly in the fields of mycology and plant pathology. It can be recorded that the collection increased from its modest beginnings to over 10,000 cultures representing more than 6,000 species by 1961. I cannot know what vicissitudes of possible inadequate support the Collection may have endured during its long and helpful history, but it is a matter of record that Dr. Westerdijk until 1952 directed its operation while she held the parallel positions of Professor of Phytopathology in the Universities of Amsterdam and Utrecht and trained fifty-six doctorates! Not until 1959 was a full-time director, in the person of Miss van Beverwijk, appointed to direct the affairs of the Centraalbureau, and I suspect that only in recent years has it enjoyed a measure of financial support reasonably commensurate with its important mission.

The American Type Culture Collection had its birth soon after Dr. C.E.A. Winslow joined the City College of New York in 1910 and became curator of public health in the American Museum of Natural History. The need for a collection of micro-organisms with special reference to public health had been recognized by the American Society of Bacteriologists and was instituted under the leadership of Professor Winslow. Later, sponsorship of the Collection was transferred to the McCormick Memorial Institute of the University of Chicago under the leadership of Professor Ludvig Hektoen. But once again it fell upon hard times and ca. 1928 it was moved to Washington, D.C. and sponsorship was vested in a Board of Trustees representing the Society of American Bacteriologists

and certain other biological societies concerned with its operations and future. Adequate housing for the Collection has posed a
critical problem and during its early years in Washington it was
moved, sometimes piecemeal, from one governmental laboratory
or educational institution to another. Finally, and perhaps in des-
peration, the Trustees voted to secure rented quarters in a former
residence and some years later to purchase a larger house on M
Street where the Collection is still located in improvised laboratories and bursting at every seam. But the Collection has finally
come of age. Aided by the active intervention and support of the
National Academy of Sciences -National Research Council more
than a decade ago, and with greater income from culture sales that
reflect the growing study and use of micro-organisms, the Collection has been able to increase steadily its accessions, its staff and
its service in recent years. Substantial grants from the National
Institutes of Health have enabled the Collection to increase its research activities and to improve its physical facilities, while a
large grant from the National Science Foundation together with
contributions from industries and other private sources insure the
erection of an adequate laboratory building on the outskirts of
Washington, for which plans have been completed and construction,
I suspect, already started. With Dr. W. A. Clark as Director and a
competent professional staff, not generously but I hope adequately
compensated, to work with him, the American Type Culture Collection after four decades of uncertainties can now look to a bright
future of productive research and service to microbiology and the
broader areas of the life sciences.

I could not possibly list the names of all individuals that have
bent an oar to bring the ATCC through troubled waters, but certain
ones stand out: Dr. Lore A. Rogers, mentor if not director, during
the transition from guest to master in its own house; Dr. R. E.
Buchanan, consistent champion of the Collection and long-time
member of its Board of Trustees; Dr. Ruth E. Gordon, Curator
who without evident sources of support held the Collection together
and enlisted the aid of NAS-NRC in its most trying hour; and
Dr. Robert D. Coghill, chemist by profession but biologist at heart,
trustee and devoted friend of ATCC who has as an individual collected substantial sums of money for the construction of its new
building.

I suspect that a comparable history, differing only in detail
from any of the above, might be related for the Algal Collection
located at Indiana University, for the National Collection of Type
Cultures in Britain, and perhaps also for others in Japan and elsewhere. But this is not my purpose. I have spoken of these three
because I am most familiar with them, and more importantly because they illustrate a point that I think is worthy of remembrance.

Namely, that each began in a modest way, and that their perpetuation and growth stemmed from the vision and selfless devotion of individuals with heavy responsibilities other than their concern for the collections per se. The picture has changed markedly in the past fifty years: these collections are now reasonably if not adequately supported, and the general public is increasingly aware of what micro-organisms may do for mankind when properly used. But we still have a job to do. As their advocates and conservators we should claim for living micro-organisms a measure of public recognition and support at least the equal of that accorded to herbaria, zoological gardens, and collections of stuffed birds for generations past.

Now that I have made this plea, I shall address myself to the topic listed in your program, although I suspect that anything I may say will be restated more competently and in greater detail in the lectures that follow relative to particular classes of micro-organisms and cell lines. Furthermore, what I can say of a general nature is perhaps already known to most of you. So, as I attempt to set the stage for the more specific discussions to follow, I hope you will forgive me if I seem to repeat the obvious in a manner that may appear trite.

Although I cannot say how and when the first attempts at preserving cultures of micro-organisms were made, I can safely conjecture that this was done by transferring them from their natural habitats to fresh and/or uninfected materials of the same kinds. Techniques were of necessity crude and sterility problematical. Emphasis soon shifted to extracts of animal tissue, plant parts, fruits and dung, oftentimes solidified after a fashion by the addition of gelatin. With the introduction of agar as a relatively inert gelling agent by Frau Hesse in the early 1880's, this was used increasingly to provide a solid base. For the most part, natural extracts continued to be incorporated as nutrients, as they still are, and the number and types of those recommended and used were and are as varied as the micro-organisms cultivated and maintained. To secure greater reproducibility, and even more to facilitate physiological and biochemical studies, attention was directed even then to culture media of known composition. As early as 1869 Raulin (24) introduced a nutrient solution containing measured amounts of sucrose, tartaric acid, and ammonium nitrate together with other salts for his study of Aspergillus niger, and subsequently Dierckx (1901) (fide Biourge (3)) modified this to secure a neutral solution that has been much used in Europe for the cultivation of saprophytic molds. A comparable medium, attributed to Czapek, which has become even more widely employed for growing the same types of fungi, was introduced in this country by Dox soon thereafter (8). Meanwhile, the bacteriologists, wherever possible, were devising

nutrient solutions of known composition to establish the nutritional
value of different carbohydrates, nitrogen sources, etc. Whether
consisting of natural extracts or mixtures compounded from the
laboratory shelf, such nutrient solutions were solidified with agar,
dispensed into tubes and as agar slants used for the transfer of
stock cultures at regular intervals as required for the specific
organisms.

This procedure, still much used and commonly termed the
agar-slant method, has certain obvious advantages. Transfers can
be made quickly, the purity and non-variability of the culture can
be verified during its development, and if stored at a relatively
low temperature (ca. 4°C) after growth (including sporulation) is
complete, it will in most cases remain viable for months or, in
some cases, even years. The culture medium, incubation temperature, and conditions of storage must of course be patterned to
minimize the production and accumulation of metabolites inim-
icable to viability. Much as can be said for defined media in
physiological studies, the use of more natural substrates often
prove advantageous for optimal growth and longevity. This is particularly true of some fungi that tend to lose their capacity to
sporulate during continued laboratory cultivation; it is even more
important in maintaining the sexual stage of many ascomycetes.
As one means of minimizing these difficulties, Snyder and Hansen
(27) have effectively used plant materials, pre-sterilized with
ethylene oxide, either incorporated within or sprinkled upon the
surface of non-nutrient agar to enhance sporulation. Whereas some
micro-organisms are amazingly self-sufficient when provided with
a few chemical compounds and thrive indefinitely, others are far
less resourceful and it behooves us to provide complex substrates
required for their maximal growth and development. The
Centraalbureau's practice of alternating substrates at successive
periodic transfers has merit for the maintenance of a large and
and diverse collection of fungi (30).

In the use of agar slant cultures renewed by periodic transfer one major hazard should perhaps be emphasized, namely the
constant threat of perpetuating some variant that under the specific
conditions employed grows more rapidly or sporulates more freely
than the more typical form one wishes to preserve. Time was when
no mycologist or plant pathologist worthy of the name would consider discussing any culture that was not of monosporous origin.
Even now we recognize this as an essential technique for determining the range of natural variability inherent in a given stock,
as demonstrated most carefully by Backus and Stauffer in their
study of Penicillium chrysogenum (2). But the problem facing the
guardian of a culture collection is not usually one of dissection;
rather it is one of maintaining in as nearly stable and reproducible

form as possible a culture with which he has previously worked or one entrusted to his care. With reference to the filamentous fungi, with which I am of course most familiar, but I suspect equally true for bacteria, yeasts and actinomycetes as well, the more prudent course is to effect a mass transfer to perpetuate what Backus has termed the "conglomerate with a pattern", if such in fact exists.

Conventional agar slant or stab cultures of micro-organisms are stored at low temperatures following growth primarily to reduce their metabolic processes, or stated differently, to retard the aging process. The same objective has been sought by other means, and prominent among these has been the practice of covering the growing culture with sterile, medicinal grade mineral oil. This was first recommended by Lumière and Chevrotier (17) for extending the longevity and reducing the variability of gonococci in serum cultures, and it has since been applied to a wide variety of micro-organisms including bacteria, yeasts, and many kinds of true fungi. Of recent years it has proved to be especially applicable for the conservation of such fungi as the Aquatic Phycomycetes that cannot be lyophilized by techniques now in vogue and the Basidiomycetes which usually consist primarily of vegetative mycelium in stock cultures. The slants must be covered completely to avoid evaporation by wicking action, and an oil overlay of approximately one centimeter in depth is generally recommended. In addition to prolonging viability, with records up to five years commonplace, in those cases where physiological competence has been tested, e.g. as regards pathogenicity, the fungi conserved under mineral oil have shown little tendency to vary. There are certain inconveniences associated with the method, but as Miss Fennell has indicated in her comprehensive review (10), it shows great promise for extending the longevity of agar-grown cultures of fungi that are not amenable to preservation by the lyophil technique.

We should perhaps mention another procedure sometimes followed by bacteriologists with apparently favorable results, but one that is generally lethal to fungi when it is applied to these highly aerobic organisms. I refer to the practice of culture storage in screw-cap tubes for the purpose of reducing possible contamination and loss of water from the agar surface by evaporation. Too often we have received for identification mold cultures of this type that appear superficially fresh and vigorous only to find that they are no longer viable. For the fungi it is far better to use a conventional cotton closure that allows free gas exchange even though the agar dries out completely.

Another method commonly employed for extending the longevity of micro-organisms consists of storage in soil or sand, a technique that came into vogue during the period from 1910 to 1920.

It was first applied to the conservation of bacteria, and the practice seems to have had a dual origin. Legume inoculants were applied to seed via soil, and as early as 1917 E. B. Fred and A. J. Riker, at the University of Wisconsin, prepared a series of stock cultures of rhizobia by mixing these in soil and storing them in Freudenreich flasks. Most of these were still viable when tested by O. N. Allen in 1946, twenty-nine years later, and some yielded typical cultures even a year ago. Parallel with this, in 1915 Speakman in Toronto is known to have stored Clostridium acetobutyticum on sand as a means of stabilizing this species (fide E. McCoy). About 1917-18 Omeliansky revived cultures of Clostridium pasteurianum from a mixture of sand and chalk left in Winogradsky's laboratory in 1894, or earlier. Based upon their experience and these latter reports, Fred and his colleagues began about 1920 to use soil routinely for the storage of bacterial stocks, principally of the groups just mentioned. In 1934 Greene and Fred (12) applied the method to a series of fungi with which they were then working, and the soil procedure then described has been adopted perhaps more generally than any other. This consists of adding one ml of a heavy suspension of conidia to 5 gm of sterile orchard loam (20% moisture), followed by drying at room temperature and storage in the refrigerator. A test last year showed Aspergillus sydowi was still viable after twenty-five years.

Other carriers including clean sand, peat soil with $CaCO_3$, and clay and sawdust have been recommended. Variations of the general method have been widely used at industrial laboratories for isolates of actinomycetes and for carrying stocks of penicillin-producing molds. Reported viabilities up to five years or more are not uncommon for a variety of fungi, and Miss Fennell (10) has cited (a) enhanced longevity, (b) minimal morphological change, and (c) ready availability of uniform inocula over long periods as advantages.

In this connection it would be interesting to have some discussion as to the manner in which soil exerts its protective influence. One can visualize this when the carrier is rich in colloidal materials, as for example garden loam or peat soil, but what can be present in sea sand to exert a beneficial effect? Is it a matter of regulating in some way the desiccation of the cells or spores that affords survival value, or, as I suspect might be the case, are the test organisms in many cases rugged types that in themselves carry the potential for unsuspected longevity. There is cause to think this may be true of Penicillium chrysogenum for tests made in Peoria several years ago showed members of this group to be almost uniformly viable after 5 years in agar slants that had been stored at temperatures often in excess of 30°C (23).

In each of the preceding methods, storage in the refrigerator is recommended as a means of limiting metabolic processes. It is not surprising, therefore, that some investigators have employed the simple device of freezing cultures and holding them in a frozen state, chipping or thawing them to remove vegetative cells or spores when sub-cultures were needed. J. W. Carmichael, of the University of Alberta, speaks enthusiastically of this method for conserving fungi (6), and Meyer (20) has recommended it as a device for preserving dermatophytes without pleomorphic change over a two-year test period. Not surprising, she found refreezing was detrimental to viability and recommends that thawed specimens be discarded. Extending the method to phytopathogens, Hamilton and Weaver (13) have effectively preserved plant tissue infected with Gymnosporangium juniperi-virgeniane and Venturia inequalis as sources of fungus inoculum for periods of 15 months by quick-freezing and subsequent storage at subzero temperatures.

With regard to cell storage in a frozen state, far more dramatic results have been obtained with glycerolized animal spermatazoa frozen at -79°C (21) and with human and animal tissue cell lines stored under liquid nitrogen. But of these techniques we shall surely learn tomorrow morning from Dr. Stulberg and others.

Many investigators have described relatively simple drying procedures as an aid to culture preservation, and these have taken a variety of forms, varying with the individual and the materials to be conserved. Fennell (10) has summarized this work as follows: (1) Rapid desiccation in vacuum, over a desiccant, or a combination of the two, is preferable to slow drying; (2) the presence of protective materials, e.g. serum or milk, improves survival; and (3) dried cultures live longer at refrigerator than at room temperatures. A few examples may be cited: Brown (4,5) preserved pneumococci and hemolytic streptococci for 4 to 12 years by drying cells in serum or blood on pieces of filter paper in evacuated bottles containing $CaCl_2$. Rhodes (25) reported 88% survival for periods up to 13 years of 61 genera of fungi by drying spores and mycelia in drops of horse serum via P_2O_5 and vacuum dehydration, followed by sealing. Cultures of the Blastocladiaceae adsorbed on strips of filter paper and slowly dehydrated were reported viable after 12 to 17 years by Goldie-Smith (11). There are, additionally, scattered reports of unusual viability of fungus cultures stored without special processing, as for example McCrea's cultures of Rhizopus nigricans at 30 years and Aspergillus oryzae at 35 years, respectively (18). Would that we knew the combination of circumstances that favor such longevities! Do they result from an ideal combination of nutrient balance and physical environment or do they reflect specific or even strain characteristics of the particular cultures?

In any case, the individual charged with the responsibility of

maintaining a substantial collection cannot place his faith in such
tenacity, and soon comes to realize that he must deal with many
variables, only a few of which are even partially understood. Of
the many different methods described and recommended, that
which combines at least two of the foregoing, namely freezing and
drying, seems to afford the greatest promise. This procedure,
commonly referred to as lyophilization, seems to have most to
recommend it for organisms to which present techniques are applicable. We shall undoubtedly hear much of this in the lectures
that follow, with many modifications designed for or found by
experience to be especially applicable to specific or more general
needs. Hence, I shall not attempt to discuss these in any detail.
Furthermore, the subject was covered in a most comprehensive
manner just two years ago by Miss Fennell in the Botanical
Review (10).

It is sufficient to say that drying cultures from the frozen
state has been practiced since the work of Shackell in 1909 (26)
and Hammer two years later (14). Limited to bacteria, viruses,
and immunological preparations for many years, it was first applied to the yeasts by Wickerham and Andreasen in 1942 (31).
Adopting their technique to the filamentous fungi, Miss Fennell and
I lyophilized the entire mold collection at the Northern Regional
Research Laboratory in 1942, while Wickerham and R. G. Benedict
(Dr. Haynes had been called by the U.S. Army) processed the
yeasts and bacteria, respectively. Results of viability tests on
selected fungi were published by us at 2 to 2-1/2 years (22), and at
6 to 7 years (9). More recently, the results of extensive tests
made on the cultures we processed in 1942 have been published by
Dr. Hesseltine and co-workers, who succeeded us at the Northern
Regional Laboratory. Of 78 species of Aspergillus and 140 species
of Penicillium, only two species of the latter genus failed to grow
when tested in 1957 (19); and of 363 cultures representing many
species in a wide variety of genera tested two years later after
17 years (16), only 8.8% were non-viable. As pointed out in our
early reports, and as confirmed by the more recent, long-term
tests of Hesseltine et al., certain fungi, particularly those with
very large and multinucleate conidia, e.g. the Entomophthorales,
are not amenable to this method of preservation. Up to the present,
the same applies generally to certain other groups, including the
aquatic phycomycetes, and strictly mycelial cultures of the higher
fungi, e.g. fleshy basidiomycetes. This is not to say that alterations in the freezing and drying schedules, the use of suspending
menstrua other than blood serum or milk, etc., might not result
in procedures that would prove applicable. In fact, this seems all
the more probable in the light of the successful lyophilization of
algae by Daily and McGuire (7) and others.

The merits of lyophilization as a method for preserving most micro-organisms is now quite well established. The many and varied reports of its successful application attest to this, as does also its general use for the conservation of stock cultures used for the manufacture of myriad products in industry. Viability is greatly extended, contamination by other micro-organisms or mites is precluded and, when properly performed, the variability among cultures resulting from lyophilized cells or spores is minimal.

I have attempted rather sketchily to indicate something of the origin of culture collections, and to relate in a general way the essential characteristics of methods developed for the care and conservation of the micro-organisms they contain. Impressive advances have been made within the past half century, and this first Specialists Conference on Culture Collections marks, I believe, the beginning of a new era in microbiology. Unmistakably, it bears witness to the quickened recognition of the opportunities and responsibilities that lie ahead for scientists interested alike in micro-organisms for what they are and for what they do. However, lest we become self-satisfied and complacent, it is well to remember that living bacteria have been reported in "wholesome canned food" (tinned meat from the Parry Arctic expedition) after 113 years (1), and that three years ago two Canadian microbiologists, Hayes and Anthony (15), reported having isolated a few micro-organisms from cores of lake sediment estimated by radio-carbon dating to be 15,000 years old. We are making progress. We also have much to learn.

References

1. Anon. The survival of bacteria in wholesome canned foods. Nature 173: 334. 1954.

2. Backus, M.P. and Stauffer, J.F. The production and selection of a family of strains in Penicillium chrysogenum. Mycologia 47: 429-463. 1955.

3. Biourgé, Ph. Les moisissures du groupe Penicillium Link. Monograph, La Cellule 33(1): 7-331. 1923.

4. Brown, J.H. The preservation of bacteria in vacuo. I. (Abstr.) Jour. Bact. 9: 8. 1925.

5. Brown, J.H. Vacuum tubes for the storage and shipment of bacteria. Science 64: 429-430. 1926.

6. Carmichael, J.W. Frozen storage for stock cultures of fungi. Mycologia 48: 378-381. 1956.

7. Daily, W.A. and McGuire, J.M. Preservation of some algal cultures by lyophilization. Butler Univ., Bot. Stud. 9: 139-143. 1954.

8. Dox, A.W. Intracellular enzymes of lower fungi, especially those of Penicillium camemberti. Jour. Biol. Chem. 6: 461-467. 1909.

9. Fennell, D.I., Raper, K.B. and Flickinger, M.H. Further
   observations on the preservation of mold cultures.
   Mycologia 42: 135-147. 1950.

10. ——————· Conservation of fungous cultures. Bot. Rev.
    26(1): 79-141. 1960.

11. Goldie-Smith, E.K. Maintenance of stock cultures of aquatic
    fungi. Jour. Eli. Mitch. Sci. Soc. 72: 158-166. 1956.

12. Greene, H.C. and Fred, E.B. Maintenance of vigorous mold
    stock cultures. Ind. & Eng. Chem. 26: 1297-1298. 1934.

13. Hamilton, J.M. and Weaver, L.O. Freezing preservation of
    fungi and fungous spores. Phytopath. 33: 612-613. 1943.

14. Hammer, B.W. A note on vacuum desiccation of bacteria.
    Jour. Med. Res. 24: 527. 1911.

15. Hayes, F.R. and Anthony, E.H. Lake water and sediment. VI.
    The standing crop of bacteria in lake sediments and its place
    in the classification of lakes. Limnol. & Oceanog. 4(3):
    299-315. 1959.

16. Hesseltine, C.W., Bradle, B.J. and Benjamin, C.R. Further
    investigations on the preservation of molds. Mycologia
    52(5): 762-774. 1960.

17. Lumiere, A. and Chevrotier, J. Sur la vitalité des cultures de
    gonocoques. Compt. Rend. Acad. Sci. (Paris) 158: 1820-1821.
    1914.

18. McCrea, A. A supplementary note on longevity of Aspergillus
    oryzae and Rhizopus nigricans. Mich. Acad. Sci., Proc. 20:
    79-80. 1935.

19. Mehrotra, B.S. and Hesseltine, C.W. Further evaluation of
    the lyophil process for the preservation of Aspergilli and
    Penicillia. Appl. Microbiol. 6: 179-183. 1958.

20. Meyer, E. The preservation of dermatophytes at subfreezing
    temperatures. Mycologia 47: 664-668. 1955.

21. Polge, C. Low-temperature storage of mammalian sperma-
    tozoa. Proc. Roy. Soc. B. 147: 498-507. 1957.

22. Raper, K.B. and Alexander, D.F. Preservation of molds by
    the lyophil process. Mycologia 37: 499-525. 1945.

23. Raper, K.B. and Thom, C. A manual of the Penicillia.
    Williams & Wilkins Co., Baltimore. 1949.

24. Raulin, J. Étude chimiques sur la vegetation des Mucedinées.
    Ann. des Sci. Nat., Bot. 5 ser., 11: 93. 1869.

25. Rhodes, M. Preservation of yeasts and fungi by desiccation.
    Trans. Brit. Mycol. Soc. 33: 35-39. 1950.

26. Shackell, L.F. An improved method of desiccation with some
    applications to biological problems. Amer. Jour. Physiol.
    24: 325-340. 1909.

27. Snyder, W. C. and Hansen, H. N. Advantages of natural media
    and environments in the culture of fungi. Phytopath. 37(6):
    420-421. 1947.

28. Thom, C. The Penicillia. Williams & Wilkins Co., Baltimore.
    1930.

29. Thom, C. and Church, M. B. The Aspergilli. Williams &
    Wilkins Co., Baltimore. 1926.

30. Westerdijk, J. On the cultivation of fungi in pure culture.
    Antonie van Leeuwenhoek, Jour. Microbiol. Serol. 12:222-
    231. 1947.

31. Wickerham, L. J. and Andreasen, A. A. The lyophil process.
    Its use in the preservation of yeasts. Wallerstein Lab.
    Comm. 5:165-169. 1942.

DISCUSSION I

by N. A. Krasilnikov
Institute of Microbiology
Moscow, U.S.S.R.

During laboratory storage micro-organisms undergo profound
changes in morphology and physiology. In the lapse of a few years
the cultures become so different as to be scarcely identifiable with
the original strains. In 2-1/2 years of growth on juice agar yeasts of
the genus Sporobolomyces become either mycelial fungi or yeasts
similar to Mycotorula and Terulopsis. The yeast, Nadsonia fulves-
vens, after five years storage on juice agar loses its morphological
distinctions and peculiar agamogenesis to be transformed into a
common yeast-like organism with limited sporulation. Nectar in-
habiting yeats of the genus Anthomyces, readily identifiable by
sight, after one year in the laboratory assume a mycelial form with
verticillate arrangement of hyphae to resemble the fungus Verticil-
lium. Profound changes were also in evidence in bacteria and
actinomycetes during laboratory storage. Thus Azotobacter
chroococcum becomes a finely dispersed culture without its characteristic intracellular inclusions, sporogenous bacteria often lose
their sporulation capacity to become asporogenous, pigmented cultures lose color, capsulated organisms are no longer able to form
a capsule, virulent strains become avirulent and active antibiotic
producers either becomes less active or inactive. It would be no
exaggeration to say that the whole variety of forms obtained experimentally using various mutagenic factors can arise spontaneously
in the process of cellular degeneration during prolonged storage in
an artificial environment.

The great mutability of cultures in storage poses a serious

problem for their preservation in the original state, not only for
industrial but also for taxonomic purposes. It is indeed essential
for a proper taxonomic analysis to have standard cultures with their
original features preserved and not storage cultures with their
specific features distorted or lost. Are there methods for the preservation of cultures which achieve this? Many different methods
for keeping cultures in storage are given in the literature (selected
media, low temperature, the anaerobiosis and dehydration of cultures and, more recently, lyophilization). Our experiments and
laboratory practice have so far yielded no reliable method for the
prolonged preservation in the original state. The best one may
hope for is the lengthening of storage life. In our laboratory routine we often use starvation media for the storage of actinomyces,
proactinomyces, mycobacteria, etc., being guided by several considerations. Degeneration proceeds faster in liquid media than on
agar media. The greater and richer the microbial mass the greater
is the mass of metabolites in the medium, and metabolites are the
mutagenic factors. In addition, the greater the mass of microbial
cells the greater is the number of deviations: variants, mutants
etc. In starvation media, such as water agar, the development of
cultures is slow and negligible, being detected only under a microscope and being limited to cellular development at the surface of
agar. Such a weak development of cultures cannot accumulate
enough metabolites to produce any harmful or mutagenic effects,
nor is the resultant microbial mass large enough to involve formation of mutants in any appreciable numbers. The same principles
remain, in a way, fundamental for some other preservation methods
as well. Thus, dried cultures do not show any sign of life, nor do
they have any metabolites and this well may be a factor in their
preservation. But the above methods of preservation should not be
applied indiscriminately to all microbial species. Actinomyces
violaceus strain 1, which produces the antibiotic mycetin, after
being kept for 30 years in conventional media (czapek agar) has
completely retained its original properties. The same strain in the
dried state in sealed ampoules is viable for no more than three
years. Under conditions of conventional storage or in starvation
media, the orange group of strains, such as A. longissimus, A.
aurantiacus, A. auroverticillatus, etc., prove rather stable. However, A. streptomycini after two to three years storage on a
czapek medium acquires clear signs of degeneration. Almost all
strains lost their ability to produce antibiotics to a certain extent.
A. coelicolor strain 8, in the same medium after two years of
storage begins to degenerate into a colourless form that does not
synthesize antibiotic. The same strain stored on water agar has
not shown signs of degeneration during seven years of storage.
Yet in the dried state, i.e. lyophilized, this strain dies in five to

TABLE I

Preservation of Actinomycetes in different media

Species	Number of strains tested	Water agar	Czapek agar	Czapek broth	Fish agar	Fish broth	Lyophilized
		Number of strains changed after 1 year storage					
A. streptomycini	120	3	24	38	48	80	5 (19)*
A. coelicolor	60	1	6	12	16	25	2 (22)
A. violaceus	10	0	0	1	2	5	0 (2)
A. longissimus	24	0	0	0	1	1	1 (2)
		Number of strains changed after 5 years storage					
A. streptomycini	80	5	28	41	38	65	14 (25)
A. coelicolor	56	3	10	23	42	51	4 (41)
A. violaceus	9	0	0	1	1	2	0 (5)
A. longissimus	21	0	0	0	2	4	2 (1)

*number of strains which died.

eight months. Different species or different strains of the same
species of actinomyces under different storage conditions differ in
their degree of stability as shown in Table I. Some of them keep
their properties well, others degenerated readily.

It is clear from the above that the cultures of actinomycetes,
as well as of other microbes, degenerate whatever the method of
preservation. To preserve the initial cultures in their original
state one has to watch them incessantly and at the first signs of
degeneration to isolate the newly formed undesirable variants from
the original cultures. This method of selection has so far been the
only one to maintain cultures in their original state since it enables
the organism to keep its morphological, physiological and biological properties intact during storage.

DISCUSSION II

by H. Proom
Wellcome Research Laboratories
Beckenham, England

My own experience has been largely confined to the operation
and maintenance of a considerable working collection of bacteria
of interest in the fields of human and veterinary medicine, and it is
to this aspect of our subject that I will limit my remarks this morning. I must confess to a feeling of affinity with Lewis Carrol's
large blue caterpillar about to make a slightly disgruntled exit
from his mushroom top in a haze of smoke having suffered somewhat in verbal argument, since in an attempt to stimulate discussion I will try and give dogmatic and simple answers to a few con-
troversial and complex questions.

It is difficult to introduce general methods of preserving
bacteria since the only acceptable method of maintaining stock
cultures is by freeze-drying. The reasons for this may not be apparent at first sight, after all, cultures of E. coli have remained
viable in sealed tubes of peptone water for twenty years, anthrax
spores in 50% glycerol saline at 4°C retain their full viable count
and virulence for periods certainly in excess of ten years, many
species remain viable for long periods stored under mineral oil
and sporulating anaerobes remain viable perhaps indefinitely in
chopped meat broth stored in the cold. These are just a few examples from a very extensive list. There are, I think, two main
reasons why these methods are not acceptable for stock cultures.
The first concerns human frailty. Any method which involves
either repeated sub-culture or periodic withdrawal of small aliquots from a container carries with it the risk and the eventual

certainty of accidental contamination. These difficulties are
reduced to a minimum with a reasonably organised freeze-drying
technique. The second reason is that these methods are less
perfect imitations of the freeze-drying technique in that the ideal
of suspended animation is less nearly approached. Methods other
than freeze-drying almost invariably involve very slow multiplica-
tion, a standard method of obtaining variants. Other difficulties
often arise, purity is a relative term, for example, with species of
anaerobes isolated from pathogenic material, it is extremely difficult to obtain absolutely pure cultures uncontaminated with unde-
tectably small numbers of the more common species. Cultures of
delicate anaerobes may be stored in meat broth in the cold room
and when examined years later give on subculture pure cultures of,
say, Cl. sporogenes, from which the original strain cannot be isolated. This is more than a theoretical objection for quite famous
collections of anaerobes have suffered badly for this reason. The
advantages of the freeze-drying technique, apart from the mundane
ones of convenience, are three-fold. The culture can be stored as
a large number of small aliquots, if one is satisfactory so are the
remainder and manipulative errors are virtually eliminated. Dried
cultures have a very long storage life varying with the species and
technique from say 5 to 5000 years, but the overriding advantage is
that desiccates retain unaltered both qualitatively and quantitatively
the characteristics of the culture dried. I believe this to be as
close an approximation to the truth as is possible in the biological
field. A negative is not subject to proof and I can only say that for
the past twenty years I have been responsible for organising and
maintaining a working collection of dried cultures. During this
period it was quite natural that any unexplained variation in growth,
toxin production, antigenicity, antigenic structure, virulence etc.
should be attributed to changes in the dried culture. However, in
every instance where it was possible to investigate this properly
the variation was shown to be due to other causes.

An example here will show how appearances can be deceptive.
A selected smooth white variant of B. polymyxa gave excellent
growth and polymyxin production from agar slope starter cultures
stored in the cold but freeze-dried cultures gave poor growth with
rough variants and little antibiotic. Clearly some change had occurred during or after drying but further investigation showed that
this was not so. It was found that the variant contained a phage
giving incomplete lysis of the smooth form only. The dried culture
was recovered in broth and the additional time in liquid culture
was responsible for this dramatic effect. The difficulty did not
arise from the drying but from the way in which it was used. To
complete the picture the culture stored on nutrient agar eventually
altered and satisfactory cultures could only be obtained by

reversion to the dried culture previously reported as completely
useless.

An understanding of the effects of cold, freezing, drying and
subsequent storage on the viability of microbial cultures is, of
course, vital to our knowledge of life processes. Perhaps relatively
little progress has been made in this field. Only partial solutions
have been found to the problems of freeze-drying bacterial cultures
for the purpose of active immunisation or starter cultures in commercial production where a minimal loss on both drying and storage at elevated temperatures are necessary requirements, but
within the context of maintaining freeze-dried cultures for reference or collection purposes there are few unsolved major technical problems. There are numerous methods with innumerable
modifications and all are more or less adequate. The difficulties
are those of organisation and selecting the technique most suitable
or convenient for the purpose in hand and also with the very few
groups not amenable to the freeze-drying process. It is difficult to
generalise about these groups but perhaps the most constant character is a less rigid cell wall. They range in difficulty from the
mycoplasma that can be maintained as desiccates but require
special care and optimal conditions to such groups as the leptospira that will not survive freeze-drying. Certainly there is scope
for technical advance here since the keeping of collections of
leptospira by constant subcultivation and repeated passage in animals is a tiresome and time-consuming business with many casual-
ties en route. The freeze-drying technique of pre-eminant status in
the bacterial field is slowly being more generally used. It is making some progress in the viral field where the method of choice is
still the deep-freeze. The problems here are somewhat different,
being more directly concerned with conditions affecting protein
denaturation. Surely it will soon become the method of choice with
a wide range of fungi and yeasts. My own limited experience with
perhaps a hundred strains of streptomyces, dermatophytes and
moulds entirely in the mycelial form would not support the fairly
widespread belief that the mycelium will not survive desiccation.
This is a point that needs clarification.

An analysis of the factors influencing survival during and
subsequent to the freeze-drying process and a more detailed discussion of freeze-drying techniques properly belongs to later
sessions and I will not mention them here. However, I would like
to indicate that there are roughly three sorts of methods, all excellent in their different ways. There is the make-shift improvi-
sation to cope with the occasional necessity in the small laboratory.
These are very effective with the tough but may break down with
the more delicate species. There are the high powered, astheti-
cally satisfying and thermodynamically efficient, centrifugal or

manifold drying processes with subsequent sealing in-vacuo.
These will give very long survival times but suffer from the dis-
advantage that drying cannot be done through bacteriological tight
cotton wool plugs and they carry a risk of contamination during
the drying process and on recovery. Finally there are methods
that involve drying under completely aseptic conditions. The one
I use is a modification of the Homer Swift technique in which the
drying is performed through a tight plug. This method gives adequate but perhaps not maximal survival times but is very suitable
for my purpose since it virtually guarantees purity and the desiccates can be used directly as starter cultures.

GENERAL DISCUSSION

E. G. D. Murray, Canada - I have over 2,000 agar-slope cultures of
a number of kinds of pathogenic bacteria, simply glass-sealed and
kept in a cupboard at ambient temperature. I have in the past
opened many such cultures after periods of sealing from a few
months to 20 years. The Enterobacteriaceae have all been alive
and unchanged in cultural and antigenic characters. I have opened
sealed cultures of other kinds and have found that many survive
well, while some, such as Pseudomonas, die out early. Leptospira biflexa was recovered after 4 years. Those I am examining
now are the shigellas that have been sealed for 44 years and
longer. So far, about 75% of these are alive and subculture readily.

R. W. Barratt, USA - We have studied the preservation of genetic
strains of Neurospora crassa by lyophil and by a new method i.e.
on anhydrous silica gel. Of about 900 cultures preserved by both
methods, we have had equal recovery in terms of viability. These
were coniidiating strains. The silica gel method has not proven
to be any more reliable for non-coniidiating strains of Neurospora
than the lyophil method. However, we believe that it will be possible, on the basis of experiments now in progress, to process
and store aconidial strains on silica gel, as well as conidial
strains.

FUNGUS CULTURES: CONSERVATION AND TAXONOMIC RESPONSIBILITY

by Emory G. Simmons
Quartermaster Culture Collection
Natick, U.S.A.

Purity, stability, and availability of cultures are major problems of all culture collections. The authenticity of their identifications, particularly as our isolates are pertinent to other scientific disciplines, should be of equally important concern.

The following remarks will (1) enumerate the most widely used methods for the preservation of fungi, their successes and shortcomings, thus serving as a basis for more detailed discussion; (2) summarize a few of the observations and maintenance problems of the primarily taxonomic laboratory with which I am associated; (3) introduce some recently developed methods; and (4) enlarge on my views on the taxonomic and informational responsibilities of a mycological culture collection.

The comments of each of us at a conference on culture collections must be weighted according to the kinds of experience he has had in the field. A brief resume of the U.S. Army's Quartermaster Culture Collection activity therefore is pertinent.

During investigations in the 1940's on the nature of microbiological deterioration of military equipment in the tropics, thousands of fungal and bacterial strains were isolated, identified, and tested for their degradative ability under controlled conditions. These organisms were preserved and became the nucleus of the Quartermaster collection, now believed to be the second largest depository of fungus isolates in the United States. The thousands of strains which have been added in the past fifteen years reflect not only a continuing basic interest in destructive saprophytic microorganisms but also the various and changing lines of research of the microbiologists, enzymologists, taxonomists, chemists, and fungicide and testing specialists who work with them.

Many of our strains are required in the microbiological degradation tests used to determine the acceptability of government purchases. Many others are used in laboratories associated with ours which have a long history of research on the enzymatic degradation of cellulose and related carbohydrates. Our primary concern, then, with physiologic stability centers around maintaining the degradative potentials and certain enzymatic capabilities of our strains.

The QM Culture Collection, like most others, has not been

particularly active in developing new methods of conservation. We
have adopted those standard techniques which have met our needs
best and which over the years have given us the most satisfactory
results, namely, periodic transfer, lyophilization, and oil-covered
slants. Of pertinence, however, are some extensive observations
we have made on a relatively large number of strains maintained
under mineral oil for a minimum of ten years. These data will be
discussed below.

Preservation Methods: Used and New

Without doubt, the commonest method of maintaining fungus
strains is as living colonies on solidified nutrient media. No organization which is working actively with fungus isolates is immune
to the time-consuming labor of transferring and rechecking many
of its strains periodically. Large collections which use no other
method of conservation face a constant technical burden and the
even more serious scientific burden of lost strains and of cultural
variation, contamination, and degeneration. Offsetting these problems is the ready availability of the cultures for comparison and
transfer.

On the other hand, a collection which maintains essentially
all of its strains only in oil-covered slants or in some other type of
relatively inactive condition cannot work rapidly or efficiently in
identification or biological survey operations. The strains may be
fairly well protected, but their utility is limited.

A compromise procedure is used in the QM Culture Collection and in several, but not all, other large collections. Each
isolate accessioned at QM is lyophilized if possible; those for any
reason unsuitable for lyophilization are stored as oil-covered
slants. In the active collection, subject to periodic transfer and
immediately available for comparison or distribution, are one or
two representative strains of each species or variety, all of the
strains derived from type material, all of the strains frequently
requested for research or developmental purposes, and all of the
strains pertinent to our taxonomic studies. Thus we maintain in
active culture 1500 strains taxonomically and physiologically
representative of the 9000 QM accessions. The inactive strains of
any species are retrieved only when needed for a new area of research, thus minimizing the attention which must be given to large
numbers of similar strains.

The biological and technical disadvantages of maintaining
cultures in an actively growing condition are grave. It is granted
that a very large proportion of any general fungus collection can be
carried through the years on a maximum of a half-dozen different
media. Thousands of strains are just as stable now as they were
when isolated years ago, and this after scores of periodic transfers
on one or two of these basic conservation media. Good examples

from our experience are two isolates which have been used very
extensively in deterioration studies, Chaetomium globosum QM 459
and a Gliocladium sp. QM 365, widely distributed and used as a
"Trichoderma sp. T-1". As nearly as we can determine, these two
strains have retained complete physiological and morphological
stability throughout 30 and 20 years, respectively, of serial transfers on potato-dextrose and hay-infusion agars. But there is an
embarrassing number of species that defy all of our best attempts
at stabilization on commonly used media. Fusarium is notorious
for its variability and its degeneration from "Normkultur"; many
phytopathogens cease to sporulate after their first transfer, thus
losing their identifiability.

For ten years my own work has centered around the taxonomy
of Alternaria, Curvularia, and Stemphylium. Cultures of a hundred
different species in these genera exhibit a discouragingly narrow
range of growth variations and color differences. Hence, stable
sporulation over a long period of years is essential to the establishment of "representative" strains. Yet many isolates of one of
the commonest pathogens, Alternaria solani, become nonsporulating after a very few transfers. It is not enough that I personally
am convinced of the identity of such isolates; they are essentially
worthless as any sort of taxonomic standard. A similar situation
obtains with Curvularia trifolii, often reported to have been isolated.
But we never have received a culture under this name that could be
reidentified; the donors of such cultures may be confident of their
original materials, but I cannot because they no longer sporulate.

Such nonsporulating strains may be usable in some kinds of
physiological work. However, a reputable culture collection which
distributes them without qualifying statements only invites disaster.
A choice must be made between (1) maintaining important non-
sporulating material without hope of reidentification and (2) putting
some of our efforts into devising means of retaining or inducing
sporulation in such strains before the character is permanently lost.

With the dark-spored genera mentioned above, diurnal light
and dark fluctuations are sufficient to induce sporulation in some
recalcitrant strains. Direct strong light, even sunlight, may be
required for others; and the extreme of direct ultra-violet irradiation is a decisive factor for yet others. Some attention has been
given to producing abundant conidia in Alternaria solani by means
of macerating liquid shake-culture pellets and then incubating them
overnight on filter paper (13). These treatments seem drastic and
troublesome as maintenance procedures and, in the case of U-V
irradiation, may result in mutations; but these or other biophysical
shock methods must be used if we are to maintain recognizable,
representative strains of many species.

Two other phenomena observed in isolates of these three

genera of Fungi Imperfecti and related to their maintenance as
active cultures should be noted, particularly as it is doubted that
these phenomena are unique to these genera. All three genera
characteristically produce relatively large, dark-walled hyphae
and multicelled conidia. When an isolate of any one of them is put
through a series of single-spore reisolations for stabilization purposes, it often happens that the mycelium gradually becomes wet,
collapsing, and pale; sporulation decreases or ceases. The culture
in this condition bears little resemblance to the original isolate.
It has been our suspicion that viruses may be the cause of such
a decline and that viral infections of fungi may in fact be rather
common. Should this prove true, we would have the explanation if
not the immediate solution for one of our conservation problems.

The second phenomenon, more readily substantiated because
it is observable under microscopic examination, is continued aber-
rant growth through many single-spore reisolations without decline
in vigor of the mycelium or of sporulation. It is possible to demonstrate very often that the hyphae (commonly $5-10\mu$ in diameter)
carry a very delicate internal hyphal parasite about 2μ in diameter.
With dilute stains these fine hyphae can be seen not only within the
host mycelium but also as minute branches extending through the
host cell walls and growing out through pores of the conidiophores
and basal scars of the conidia. They are of an essentially different
magnitude, color, and degree of fragility than those of the host's
branches or germination hyphae. We have never been successful in
establishing cultures from hyphal tips of these internal parasites.

It is asserted, then, that in addition to variation exhibited in
genetically unstable but otherwise truly pure cultures, we are faced
with the necessity of combating variation and instability incited by
internal fungal and possibly viral parasites. Neither of these latter
areas of variability has been studied in any degree; much basic
work in them is needed and should be both enlightening and profitable.

The QM active collection is transferred twice each year; the
cultures are maintained in refrigerators at about 5°C between the
semiannual transfers. Remarkably few strains are lost through
such handling. Commonly we must retrieve 50-75 strains from
lyophilized or oil-covered preparations after each full transfer of
cultures. The strains which fail to grow at any given transfer
seldom are the same ones year after year.

For active maintenance of each culture we normally use the
same medium indefinitely after it has been chosen. The choice is
made after reisolation and study of each new strain on several
different media; the conservation medium chosen is the one on which
the isolate is culturally most stable and best sporulating. Czapek
and malt agars are used for most of the Aspergillus, Penicillium,

and closely related strains. Potato-dextrose agar still is used for a considerable portion of our cultures which have remained stable on this relatively rich medium. Gradually we have been shifting great numbers of isolates, particularly new ones, to media in which the only nutrients are plant juices or decoctions. "V-8 juice" (Campbell Soup Co.) is an excellent substrate for a wide range of saprophytes and plant pathogens such as those found in Stemphylium and Curvularia. Growth is rapid and abundant on 20% V-8 agar (16), and sporulation often is phenomenal. Potato-carrot agar (12; 5), based on a very weak decoction of these two vegetables, gradually is being used to replace PDA for strains which find this latter medium too rich for good sporulation.

One other medium, hay-infusion agar (21), deserves special mention and recommendation, both for isolation work and for conservation procedures. Hay agar is based on a weak decoction of dead, partially decomposed roadside grasses. The medium is completely undefined in a chemical sense except that it is used at pH 6.0-6.5. Its most helpful characteristic is that it does not support the development of masses of mycelium which often interfere with isolation work but that it does permit the abundant and typical sporulation of great numbers of both pathogenic and saprophytic molds. It has proved most helpful in isolation of dark-spored Hyphomycetes, particularly Alternaria, which tend to be heavily mycelial on any medium with added sugars; it is the conservation medium of choice for any Alternaria or Curvularia which will sporulate on it. This same characteristic of suppressing mycelium in favor of sporulation is particularly welcome in the lyophilization procedure where large numbers of spores and relatively little mycelium are desired.

The value of lyophilization for conservation of fungus strains is best documented by the series of reports issued since 1945 (17; 7; 14; 8) by mycologists of the U.S.D.A. Northern Regional Research Laboratory. NRRL pioneered this technique for fungi in 1942 and has firmly established its utility on the basis of intensive experience. Most of the large public and industrial collections in the U.S. have adopted lyophilization procedures, QM having done so in 1944. The observations of all these laboratories are roughly the same, namely, that well over 90 per cent of fungus strains which produce spores can be preserved in lyophil for periods of at least 10-15 years; the viability expectation is much greater than this.

It is a common observation that lyophilized cultures retain their morphological characteristics at least as well as the same strain maintained by periodic transfers. Our laboratory and others (8) have observed that many isolates retrieved from lyophil are morphologically more stable and sporulate more profusely than the same strains which have been carried actively on nutrient media. In particular, isolates which show a tendency to have nonsporulating patches occasionally stabilize quite well and sporulate vigorously when retrieved from lyophil. It is believed that the mycelial patches are genetically variant and that, because they are nonsporulating, they are eliminated under the rigors of the lyophilization process.

One area of failure with the lyophilization technique deserves special mention and further attention. Neither NRRL (8) nor QM has had any degree of success in maintaining species of Entomophthorales in lyophil. Although some strains, particularly one of Delacroixia coronata (QM 6844, NRRL 1912), produce spores abundantly, only a very poor degree of viability has been achieved after freeze-drying. It has not yet proved possible to extend even this minimal viability for longer than five years, most preparations proving to be nonviable immediately after processing.

References to the retention of physiological characteristics by lyophilized fungi are meager. Our own observations on the retention of cellulolytic activity reveal that this character has not changed appreciably in fungus strains which were culturally stable when lyophilized. Dorothy I. Fennell (6) has made the pertinent remark that "the widespread use of the lyophil method of preservation by industrial laboratories, where vast and expensive fermentations depend on the stability of their cultures, seems to offer overwhelming evidence that physiological characters are maintained unchanged by the great majority of lyophilized cultures."

Preservation of cultures for relatively long periods of time under a layer of sterile mineral oil is a method which has many users but few advocates if any other means of maintenance can be applied. The technique is used routinely for the storage of large numbers of Basidiomycete cultures. The Division of Forest Products, C.S.I.R.O., Melbourne, maintains its collection of 1400 strains in this manner with subculturing every three years. They report better than 99 percent viability with no detectable change in physiologic characteristics, although some of the same strains maintained as active cultures have lost their ability to decay wood (18). The major application of this technique at QM is only to mycelial or poorly-sporulating strains which are not amenable to lyophilization.

In the early development of the QM Culture Collection, a total of 2000 fungus cultures representing a wide range of tropical isolates were prepared as oil-covered slants. With the exception of a few Phycomycetes, all other isolates selected as representative of the collection remained viable for 12-24 months (1). In 1955-1956, ten years after these slants were covered, all of the cultures were checked for viability. Of the original 2000 cultures (some duplication) 669 were viable (Table I).

TABLE I*

Ten-year viability of cultures maintained under oil

		Viable	Dead	% Viable
Actinomycetales	Actinomycetaceae	4	13	24
Mucorales	Mucoraceae	10	10	50
	Mortierellaceae	0	1	0
	Piptocephalidaceae	6	2	75
	Choanephoraceae	1	38	3
Phycomycetes (Unidentified)		0	3	0
Endomycetales	Endomycetaceae	0	2	0
Eurotiales	Gymnoascaceae	2	1	67
Sphaeriales	Sphaeriaceae	53	9	85
	(Chaetomium)	(52)	(6)	
	Hypocreaceae	2	5	29
Basidiomycetes (Unidentified)		15	36	29
Moniliales	Pseudosaccharo-mycetaceae	4	1	80
	Moniliaceae	38	126	23
	(Gliocladium)	(1)	(44)	
	Dematiaceae	161	160	50
	(Alternaria)	(25)	(2)	
	(Cladosporium)	(47)	(16)	
	(Curvularia)	(45)	(9)	
	Stilbaceae	1	4	20
	Tuberculariaceae	92	312	23
	(Fusarium)	(82)	(285)	
Melanconiales	Melanconiaceae	5	41	11
	(Pestalotia)	(2)	(39)	
Sphaeropsidales	Sphaerioidaceae	189	159	54
	(Phoma)	(68)	(12)	
	Nectrioidaceae	2	3	40
	Leptostromataceae	0	1	0
Mycelia Sterilia		79	329	19
Unidentified		5	33	13
		669	1289	34

*Prepared by Miss Dorothy I. Fennell.

Because of its simplicity, the oil-cover technique will remain useful for relatively small general collections. We do not yet have a good substitute for-moderately long-term conservation of non-sporulating isolates. Special attention in any case must be given to the quality of the oil, to its initial sterility and dryness, to the life-expectancy of individual strains under these conditions, and to proper storage temperature for particularly delicate organisms.

With a limited staff and a large collection, periodic transfer, lyophilization, and oil-cover techniques are the most generally useful. Two other techniques, still under observation and development, eventually may replace or supplement the older standard methods.

Storage of cultures at -18° to -20°C ("deep freeze storage") is gaining attention. One general collection of 400 cultures has been stored thus for nine months on Sabouraud's agar in screw-cap tubes (2). In general, viability was maintained well through this period for a great variety of species, including strains of the usually cold-sensitive Choanephora as well as numerous isolates of dermatophytes. A second more extensive report is based on storage of 451 isolates for five years on PDA in screw-cap tubes (11). A total of 331 isolates (228 species) remained viable, these including a wide range of saprophytes and dermatophytes. All but 55 isolates, which included several of Curvularia and Helminthosporium, survived for at least two years in frozen storage. The results indicate that, after testing the survival capacity of individual isolates, deep-freeze storage of a broad spectrum of fungus cultures is a reliable procedure if new transfers are prepared every two years.

The most recent extension of preservation by freezing involves maintaining sealed vials of mycelium and spores suspended in 10% glycerol at "ultra-low" temperatures, specifically at -184.5°C in a refrigerator cooled by liquid N_2 (9). The organisms chosen for this study were ones which previously had not survived lyophilization. Included were 12 strains of Rhizoctonia, Botrytis, Pythium, Phytophthora, Syzygites, and Choanephora. All survived this drastic treatment for three days. It remains to be seen whether or not this method can be applied to a wide range of species and whether it will permit retention of viability over long periods of time.

Many laboratories, primarily industrial ones, at one time or another have used some method of preserving spore populations in sterilized soil or sand. The method has proved particularly acceptable for maintaining strains of Fusarium (15) and other equally mutable organisms (10; 3) in a stable genetic condition. With the advent of freeze-dry procedures, however, maintenance of cultures in sterile soil would appear to have lost most of its advantages, with the one exception of serving as a ready supply of abundant uniform inoculum over long periods of time.

Conservation of fungus strains on plant parts, especially leaves, is a classic technique for plant pathogens such as the

Uredinales. A report of the continued isolation of Helminthosporium
sativum from wheat seed after 17 years of storage (19) is remarkable and suggestive for development, particularly as species of
parasitic Helminthosporium have a poor history of survival in lyophil or under deep-freeze conditions (11). Successful establishment of a Rhizoctonia in seeds of 15 varieties of grains and of
Helminthosporium carbonum in wheat seed is leading to attempts
to extend this technique to numerous other fungi (4). If completely
non-infested seed stocks can be obtained, maintenance of pure cultures in the form of artificially infected seeds will be of high value.

Taxonomic and Informational Responsibilities

Some microbiologists are interested only in what a specific
fungus strain can do for them. Its Latin name is a bothersome
anachronism and its strain number is useful primarily for listing
in published tables and in requesting replacement of contaminated
cultures. But such workers unwittingly close the door on one of
their most potent tools, namely, the published literature to which
the key is proper identification of the pertinent micro-organisms.
In our time one mutant of one strain of one isolate of one fungus
species may have achieved fantastic research and industrial importance; this living entity then becomes a precisely engineered
tool with a special usefulness just so long as it retains its biosynthetic ability.

On the other hand, the microbiologist who is looking for a
better or more versatile tool for biochemical studies or for industrial yields instantly recognizes the utility of a carefully classified
organism. The taxonomy of the species immediately suggests to
him related strains or species worthy of biochemical investigation.
F. H. Stodola, in praising his mycological colleagues at NRRL, has
commented (20): "the present high yields of penicillin, riboflavin,
dextran, vitamin B_{12} and the carotenoids can be traced directly to
suggestions from taxonomists that certain taxonomic groups ought
to be good places to look for the desired organisms."

The point is this: certainly, as specialists in the handling of
fungus cultures, we must do everything possible to insure the
purity, the morphological and biochemical stability, and the availability of our isolates throughout the years; but just as certainly,
we must assume with a good will the triple responsibilities (1) of
identifying our own isolates with all possible precision, (2) of
checking the identifications of isolates we accept from others, and
(3) of contradicting shoddy or irresponsible cultural and taxonomic
work. To do less than this is to bring confusion into the scientific
disciplines that use our cultures and to invite ill-repute to ourselves.

QM interests in cultures, conservation, and classification
have been extended in recent years to the development of a

National Index of Fungus Cultures for isolates maintained in laboratories throughout the United States. It is a common observation that over the years, as one microbiological problem after another yields pertinent data, a mass of information (often unpublished) accumulates on specific strains. Such data remain of restricted interest and of little use to anyone except the person who developed them. This situation exists in hundreds of culture collections, large and small, specialized and general. As the Microbiological Era has developed during the past 20 years, more and more often the following questions are being asked: is there a fungus strain which can perform such-and-such a biological task; and where can I obtain such an organism? If such an isolate exists in one of the large collections of the world, we are likely to know about it. But if such a strain is held in a small university or other research laboratory, the fact may escape us for years.

The National Index of Fungus Cultures was organized in our laboratory in 1955 to develop as a centralized source of as much information as can be extricated from the records of cooperating collections in the U.S. The bulk of the data centers around biological characteristics and the special usefulness of specific isolates. Additional information includes identity, authenticity, pathogenicity, special culture requirements, viability in culture, and isolation history. Each item of information, suitably coded, enters an electronically operated punched-card system. The N.I.F.C. then functions on request as a source of specialized information on the characteristics, location, and availability of these cultures. The program, although of special interest to our laboratories, has been designed and will function without restriction in the service of all microbiological disciplines.

References

1. Buell, C.B. and Weston, W.H. Application of the mineral oil conservation method to maintaining collections of fungous cultures. Am. Jour. Bot. 34: 555-561. 1947.
2. Carmichael, J.W. Frozen storage for stock cultures of fungi. Mycologia 48: 378-381. 1956.
3. Ciferri, R. and Redaelli, R. Mancata formazione di forme ascofore e conservazione di culture de funghi patogeni in substrati naturali. Mycopath. 4: 131-136. 1948.
4. Crosier, W.F. Storage of disease fungi in seeds. Farm Research 27: 10. 1961.
5. Dade, H.A. Laboratory methods in use in the Culture Collection, C.M.I., in Herb.I.M.I. Handbook. Commonwealth Mycological Institute, Kew. 1960.
6. Fennell, D.I. Conservation of fungous cultures. Bot. Rev. 26: 79-141. 1960.

7. Fennell, D. I., Raper, K. B. and Flickinger, M. H. Further investigations on the preservation of mold cultures. Mycologia 42: 135-147. 1950.

8. Hesseltine, C. W., Bradle, B. J. and Benjamin, C. R. Further investigations on the preservation of molds. Mycologia 52(1960): 762-774. 1961.

9. Hwang, S. W. Effects of ultra-low temperatures on the viability of selected fungus strains. Mycologia 52(1960):527-529. 1961.

10. Jones, K. L. Further notes on variation in certain saprophytic actinomycetes. Jour. Bact. 51:211-216. 1946.

11. Kramer, C. L. and Mix, A. J. Deep freeze storage of fungus cultures. Trans. Kans. Acad. Sci. 60:58-64. 1957.

12. Langeron, M. Précis de Mycologie. Masson & Cie, Paris. 1945.

13. Lukens, R. J. Conidial production from filter paper cultures of Helminthosporium vagans and Alternaria solani. Phytopath. 50: 867-868. 1960.

14. Mehrotra, B. S. and Hesseltine, C. W. Further evaluation of the lyophil process for the preservation of Aspergilli and Penicillia. Appl. Microbiol. 6: 179-183. 1958.

15. Miller, J. J. Cultural and taxonomic studies on certain Fusaria. I. Mutation in culture. Canad. Jour. Res. C. 24: 188-212. 1946.

16. Miller, P. M. V-8 juice agar as a general-purpose medium for fungi and bacteria. Phytopath. 45: 461-462. 1955.

17. Raper, K. B. and Alexander, D. F. Preservation of molds by the lyophil process. Mycologia 37: 499-525. 1945.

18. Rudman, P. (Letter of information re this Conference; C.S.I.R.O., Melbourne; August 1961.)

19. Russell, R. C. Longevity studies with wheat seed and certain seedborne fungi. Canad. Jour. Pl. Sci. 38: 29-33. 1958.

20. Stodola, F. H. The value of taxonomy and culture collections in fermentation research. (Comments at Pasteur Fermentation Cent.; New York, November 1957.)

21. Thom, C. and Raper, K. B. A manual of the Aspergilli. Williams & Wilkins Co., Baltimore. 1945.

DISCUSSION I

by B. L. Brady
National Collection of Yeast Cultures
Brewing Industry Research Foundation
Nutfield, England

The National Collection of Yeast Cultures is responsible for maintaining stock cultures of yeasts of industrial and academic interest in Britain. Although all kinds of nonpathogenic yeasts come within the scope of the collection, about one quarter of the 800 different strains maintained consists of top fermentation brewing yeasts banked for the members of the British brewing industry.

Morphologically the cultures maintained, being yeasts, show less diversity than a collection of fungus cultures, and perhaps as a result of this are less difficult to maintain. We use the three main methods that are used in the Quartermaster Collection; periodic transfer, lyophilization, and maintenance on oil covered slants.

Like most organizations where a large amount of routine work has to be done by a limited number of workers, a compromise has to be achieved between what is theoretically desirable and what is practically possible. Thus although a certain amount of experimental work has been carried out on methods of preservation and on variation of strains in culture, most of our efforts are directed towards preserving the yeasts by tried methods. A regular check is kept on the morphological and biochemical characters of all the strains, a suitable "spot" test being chosen in each instance.

Pure cultures

Yeasts deposited with the Collection are always checked for purity when they first arrive, by streaking on malt extract agar, selecting a single colony and restreaking to a total of three times, before stock cultures are made. If, during the course of streaking morphological variants are observed, these are kept separately in the Collection, both as growing cultures and in the freeze-dried state. In this way we attempt to compromise between single cell isolation where characteristic variants may be lost and mass propagation where contaminants may be concealed. Even so, we have the occasional strain which habitually produces "rough and smooth" or "pigmented and non-pigmented" variants from a stock culture originally started from a single colony. It is evident that work on the incidence and causes of clonal variation within strains would be useful in this connection.

Variation

One of the reasons for keeping our stock cultures in liquid medium is in order to discourage the incidence of sporulation and consequent possible genetical variation. We have, in fact, recorded very little variation in the cultures which in some instances have been kept for over 40 years without apparent change in their properties. Occasionally "deterioration" in a certain property is recorded for those yeasts kept in liquid medium, these properties including spore-forming ability, the production of a pseudomycelium and in the formation of a "starch" capsule, similar changes as are noted by Wickerham (8). Likewise an occasional loss or diminution of a biochemical character is noted particularly in the ability to

ferment maltose or melibiose. In some few instances a strain
which has shown a requirement for a certain amino acid when
deposited has lost this requirement on repeated culture.

Methods of preservation

Yeast cultures are readily preserved in an active condition
in malt extract medium, subject to the occasional disadvantages
of variation mentioned above. The medium used for the National
Collection of Yeast Cultures is Wickerham's malt-extract, yeast-
extract glucose peptone ("M.Y.G.P.") (8) which is a generally useful
medium, supporting those yeasts which utilize glucose only as a
carbon source as well as those capable of using the sugars present
in malt. A few more exacting strains are kept on specialized
media. The yeasts are kept at 25°C until past the logarithmic
phase of growth and are then stored at approximately 4°C. Slant
cultures on a similar medium containing agar are maintained at
20°C. The whole collection is regularly sub-cultured twice yearly.

Of the methods for maintaining cultures in a semi-active
condition, the paraffin oil covered agar slant is not, in our experience, of such general use for preserving yeasts as it has been
shown to be for other fungi, but it is a useful ancillary method.
Some collections of yeasts are successfully preserved in 10%
sucrose solution with no nitrogen source present. The smallest
possible inoculum is used and the culture maintained at 0°C
whereby it is kept in a state of inactivity with the minimum of
reproduction by budding (3). Another such method which is being
increasingly used for bacteria is direct freezing, frozen cultures
being stored in a deep freeze around -20°C to -40°C. This had
been used successfully for fungi and certain yeasts (2; 7) although
at the period of publication these cultures had not been checked
for viability after a period longer than 1 year from first freezing.
A considerable amount of work has recently been done on the
optimum conditions for this method using bacterial cultures and
it is probable that some of the variations suggested might be
applicable to yeast and fungus cultures (6).

Turning to lyophilization, this is a generally satisfactory
method of maintaining yeasts in an inactive condition over long
periods and consequently reducing the risk of mutation during
growth. As most yeasts approach the unicellular condition, the
difficulty often encountered with fungi that only actively sporulating cultures survive lyophilization does not arise, although it is
apparent that those yeasts which exist largely as mycelium survive freeze-drying least well. In our experience, yeasts belonging
to certain genera show in general a higher viability immediately
after freeze-drying than those from other genera, Rhodotorula and
Debaryomyces surviving well, Saccharomyces poorly (1). In many
instances a parallel appears to exist between species that survive

in and can be isolated from the atmosphere and those which survive
lyophilization. Further, as might be expected, those yeasts which
have comparatively large cells, like Saccharomyces cerevisiae
and Saccharomycodes ludwigii show a much lower viability on
freeze-drying than those with smaller cells, and the fact that most
of the known haploid strains we have lyophilized, including those
derived from Saccharomyces cerevisiae, show a high initial
viability, may be a reflection of cell volume.

Much work on the fundamental aspects of freeze-drying has
been published recently, mainly in connection with bacteria, but
Mazur (4) and Nei (5) in particular have been concerned with S.
cerevisiae. It is to be hoped that some of the information obtained
from this work may advantageously be used to adapt the routine
methods of preservation of collections of fungi and yeasts.

References

1. Brady, B. L. Some observations on the freeze-drying of yeasts.
    In Recent research in freezing and drying. Edited by
    A. S. Parkes and Audrey U. Smith. Blackwell Scientific
    Publications, Oxford. 1960.
2. Carmichael, J. W. Frozen storage for stock cultures of fungi.
    Mycologia 48: 378-381. 1956.
3. Devreux, A. Levains et fermentation. Fermentatio 5: 289-293.
    1959.
4. Mazur, P. The effects of subzero temperatures on micro-
    organisms. In Recent research in freezing and drying.
    Edited by A. S. Parkes and Audrey U. Smith. Blackwell
    Scientific Publications, Oxford. 1960.
5. Nei, T. Effects of freezing and freeze-drying on micro-
    organisms. In Recent research in freezing and drying.
    Edited by A. S. Parkes and Audrey U. Smith. Blackwell
    Scientific Publications, Oxford. 1960.
6. Postgate, J. R. and Hunter, J. R. On the survival of frozen
    bacteria. J. Gen. Microbiol. 26: 367-378. 1961.
7. Tanguay, A. E. Preservation of microbiological assay orga-
    nisms by direct freezing. Appl. Microbiology 7: 84-88. 1959.
8. Wickerham, L. J. Taxonomy of yeasts. Tech. Bull. No. 1029.
    United States Dept. Agric., Washington, D. C. 1951.

DISCUSSION II

by Edward O. Stapley
Merck Sharp & Dohme Research Laboratories
Rahway, U. S. A.

From the standpoint of the industrial microbiologist, the
problem of culture preservation is an extremely important one. In

general, we are not as concerned over the permanent storage problem as with what may be termed "the interim period". Those cultures which have been selected because of some interesting property are usually preserved as lyophilized stocks. Once such a lyophilized stock has been proven by test and found satisfactory, no further problem is usually encountered. The time interval between primary test of an isolate and the point when a decision with respect to future interest in the culture can be made is the critical period.

   To illustrate the magnitude of the problem, let us consider a theoretical screening program. Fungi are isolated and tested at the rate of 1000 per week with 10 per cent having some property (such as antibiotic activity) which is of interest. When tests to evaluate the culture's activity require several weeks, thousands of isolates accumulate before final decisions are reached. The question of how to maintain such cultures is one of our chief concerns.

   The rapid loss of interesting properties of fungi isolated from soil has been a serious problem. The data in Table I reveal that, in a typical screening program for production of antibiotics, only 29 per cent of the active isolates repeated their activity when retested. Many reasons may be advanced for this lack of reproducibility; however, loss of the original capacity as a result of selective pressures during growth and storage is certainly an important contributing factor.

TABLE I

Antibiotic production of fungi isolated from soil

	No. tested	No. active	% active
Primary test	6, 105	741	12
Secondary test	644	186	29

The relationship between degree of activity and reproducibility is summarized in Table II. It is obvious that cultures with borderline activities do not hold up well on retest. Nevertheless, it is of interest to note that almost half of the isolates with high initial activity failed at the secondary stage. This indicates that the apparent loss of activity is only partly the result of quantitative variation.

TABLE II

Antibiotic production of fungi

Relationship of magnitude of activity to reproducibility

Initial trial	No. retested	Active No.	Active Per Cent
Low activity	56	6	11
Average activity	32	16	50
High activity	14	8	57

The likelihood of appearance of activity in subsequent retests is considered for an antitumor screening program in Table III. In this case the active fungi were regrown four times. Twenty-nine per cent of the cultures were not active in any of the trials and 33 per cent repeated activity only once. Thus, it is apparent that the interim storage problem applies to more than one type of activity and cannot be explained away on the basis of variations in conditions or tests as repeated trials have been observed to fail.

TABLE III
Reproducibility of antitumor activity of fungi

Result of four retests	No.	Per Cent
Active in four trials	15	9
Active in three trials	17	10
Active in two trials	33	19
Active in one trial	59	33
Not active	51	29
Total	175	100

The qualitative aspect of this problem is the most serious consideration. If certain characteristics are more ephemeral than others, it is likely that cultures with interesting biochemical characteristics are being repeatedly overlooked because of inadequate methods of preservation.

One is inclined to take issue with the concept that ready reference requires the maintenance of a large stock of cultures in the form of vegetative slants. As a result of such a procedure, where slant-to-slant transfer is practiced, many characteristics of micro-organisms may be expected to change. In our experience, the most satisfactory procedure for preservation has been lyophilization. Cultures used repeatedly are transferred periodically, as needed, from lyophilized stocks to a number of replicate slants. The number of such replicates varies with the individual program. Some of the original lyophilized stock should always be maintained, so that replacements can be prepared from the original permanent storage form of the culture. Such a procedure minimizes the amount of variation which can occur in stock cultures.

The maintenance of taxonomic characteristics is an important consideration. It must be assumed that cultures which have lost some of their typical properties, such as sporulation, have also lost or gained other properties. Thus, biochemical properties of atypical cultures cannot be considered necessarily representative for a particular species. On the other hand, taxonomic identity is not a proof of genetic homogeneity. Statistically, we must expect a slant culture of billions of cells to contain many genetic variants. Any set of isolation and vegetative maintenance conditions will exert a selective pressure. A priori, we have no means of assurance that the selection will be favorable to our interests.

Culture preservation starts with culture isolation. In the absence of definitive studies on the effect of methods of storage for the "interim period", we are forced to rely upon minimizing the vegetative passage of new isolates as the only method of reducing variation.

GENERAL DISCUSSION

H. Katznelson, Canada - Dr. Brady, under what conditions would you recommend storage of osmophilic yeasts?

Miss Brady, England - We have found that they have survived very well when lyophilized in the same way as the other collection, i.e. growing them in malt extract medium and freeze-drying them in vacuum over phosphorous pentoxide in small ampoules.

Miss Deitz, USA - I would like to mention that I have maintained all our organisms -actinomycetes, fungi and yeasts - on soil for at least 8 years with no loss of viability. We are also maintaining a few algal cultures in this way.

E. G. Simmons, USA - Most of the discussions so far have had to do with organisms that produce resistant bodies of some sort - conidia, spores, etc. There are, of course, among the fungi many isolates in which there are no resistant bodies of this sort. I have been asked several times how to maintain the characteristics of non-sporulating basidiomycetes in culture and in this regard I have talked with Dr. Mildred Nobles, of Ottawa. She uses a very weak malt extract agar (1 1/2% malt) with no other additives and has had little or no trouble with deterioration in her basidiomycete cultures, either in their morphology or in their destructive capacities.

Turning to liquid nitrogen storage, I would like to ask Dr. Hwang if she has had any true failures with maintenance of fungi at liquid nitrogen temperatures.

Miss Hwang, USA - So far we have had no failures, much to our surprise. The fungi we have tried are mostly asporogenous or the ones which cannot be freeze-dried.

G. Lindeberg, Norway - We have experienced the loss in ability to form one specific enzyme in a culture kept on malt agar. This was with a strain of Marasmius sclorodonius which we found produced tyrosinase. Some two years later the strain had lost its ability to form this enzyme whereas it still produced laccase, which is always found in this fungus.

T. O. Wikén, Netherlands - The higher basidiomycetes are now of considerable industrial importance. As we work out methods for cultivating and preserving mushroom and other basidiomycetes and higher ascomycetes, I think that they will become even more important, because of the extracellular substances which are produced and because of the high content of protein and lipid.

MAINTENANCE OF STRAINS OF BACILLUS SPECIES

by R. E. Gordon and T. K. Rynearson
Institute of Microbiology, Rutgers, the State University
New Brunswick, U.S.A.

Introduction

Preservation of strains of Bacillus species is comparatively easy provided the cultures form spores, because the spores remain viable for many, many years. Factors affecting sporulation by the bacilli, subject of much study, were ably reviewed by Knaysi (10,11) Curran (4), and Ordal (14) in connection with their own work. Later pertinent papers were those of Campbell and Sniff (2); Kanai (8); Tomcsik, Bouille, and Baumann-Grace (20); Gollakota and Halverson (6); Ierusalimsky and Egorova (7); Long and Williams (13); Shinagawa (16); Brady, Chan, and Pelczar (1); Fields (5) and Kolodzie and Slepecky (9). These studies indicated that spore formation was affected by many nutritional and other environmental factors and that mineral salts were important. Nearly all the studies of the last 20 years, however, were limited to a few strains of one or two species. Three exceptions to this limitation were: Charney, Fischer, and Hegarty (3), who examined nine strains representing seven species and reported that manganese was essential for sporulation by the cultures of six of the seven; Schmidt (15), who found variation in sporulation by seven flat sour obligate thermophiles in response to different concentrations of nutrients and salts; and Tomscik, Bouille, and Baumann-Grace (20), who, after a study of 35 strains of B. cereus, one of B. thuringiensis, and 12 of B. anthracis, concluded that sporulation on several media was influenced by strain variation.

A taxonomic study of the genus Bacillus, reported by Smith, Gordon, and Clark (17), and a recent examination of the American Type Culture Collection's strains of Bacillus spp. demonstrated the value of soil extract in maintaining cultures of many of the species. Observations on spore formation by approximately 1500 strains, the media, and methods used in preserving the strains are presented here.

Material and Methods

The numbers of strains of Bacillus spp. maintained and examined are listed in Table I. Strains of B. cereus include those of varieties mycoides, anthracis, and thuringiensis.

Soil extract. Garden soil, rich in organic material, was air-dried by spreading it in a thin layer, crushing it, and stirring it at intervals. When dry, the soil was sifted through a coarse sieve. Four hundred gm were mixed with 960 ml of tap water in a 2-liter flask and autoclaved at 121° for 1 hour. Several flasks of soil

PLATE 1

Soil extract agar

Nutrient agar

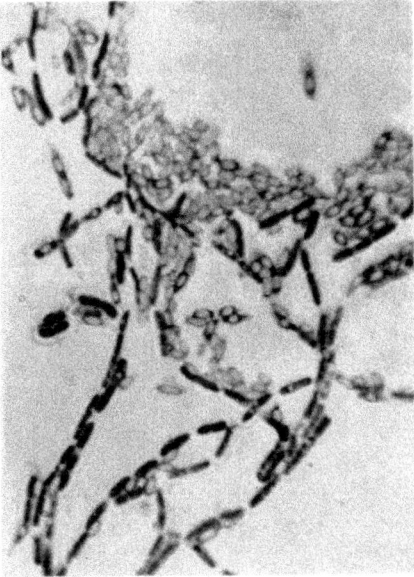

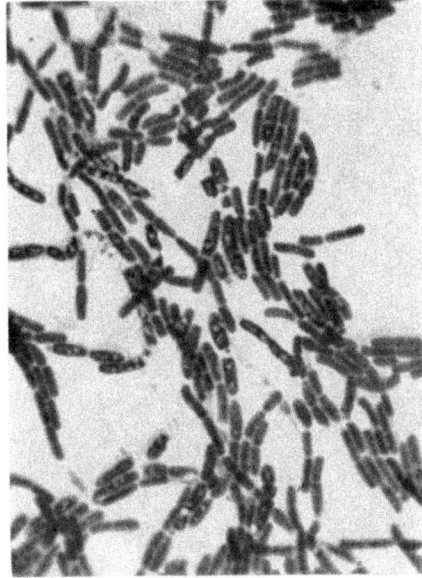

B. cereus 10206

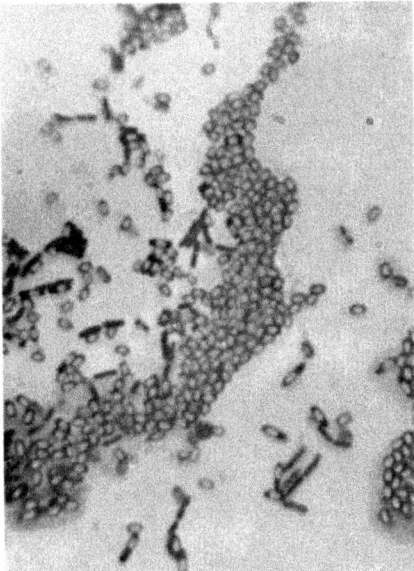

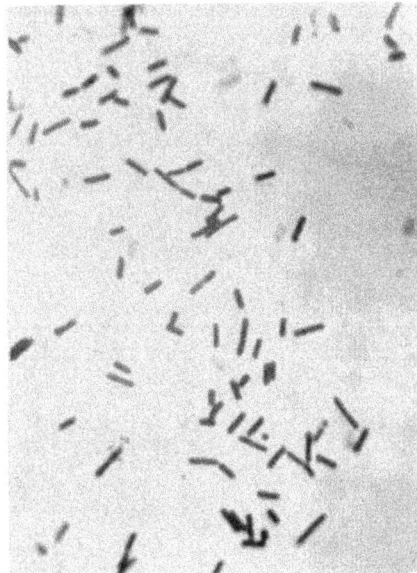

B. subtilis 6051

PLATE 2

Soil extract agar   Nutrient agar

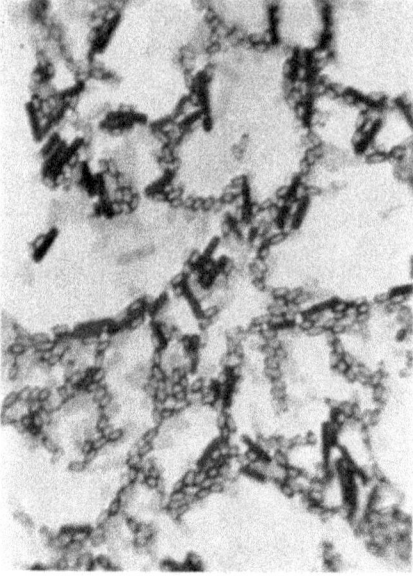

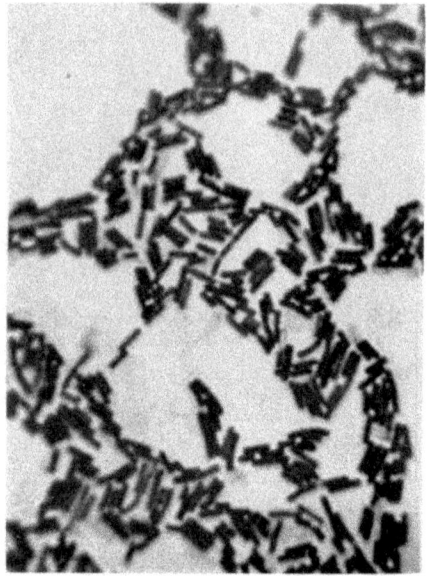

B. badius 14574

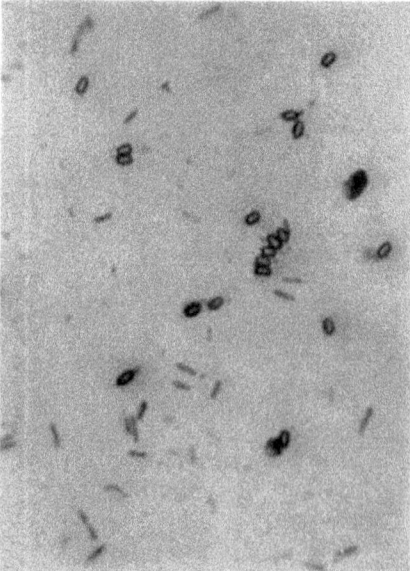

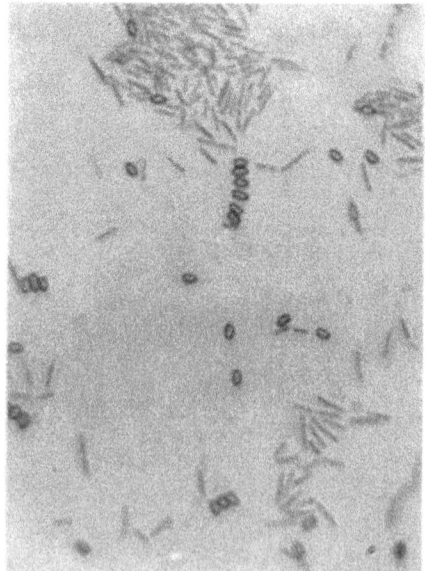

B. alvei 10871

PLATE 3

Soil extract agar

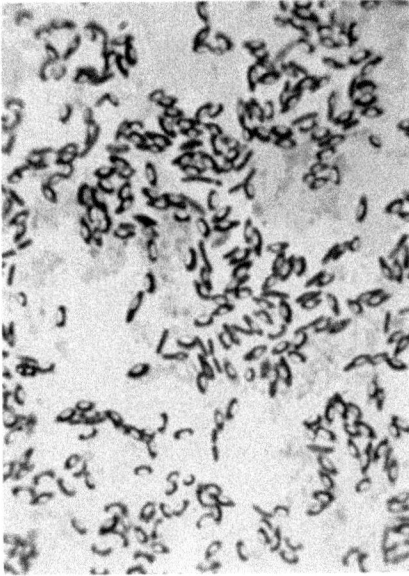

Nutrient agar

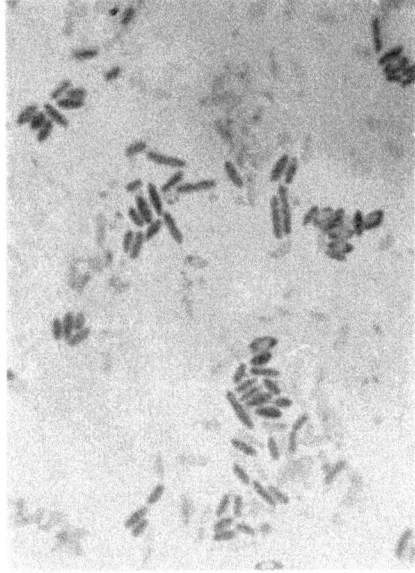

B. laterosporus 9141

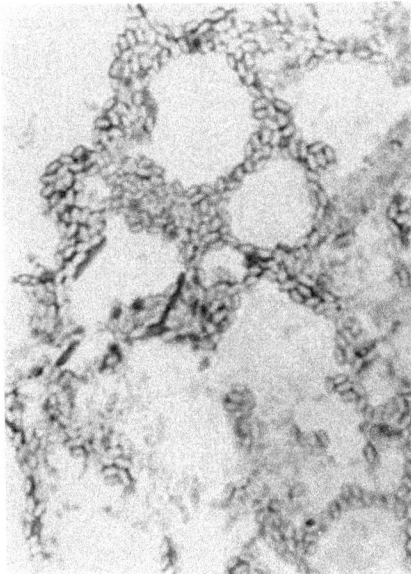

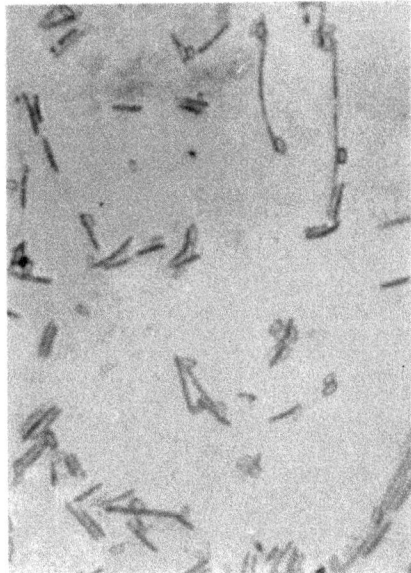

B. brevis 10027

PLATE 4

Soil extract agar   Nutrient agar

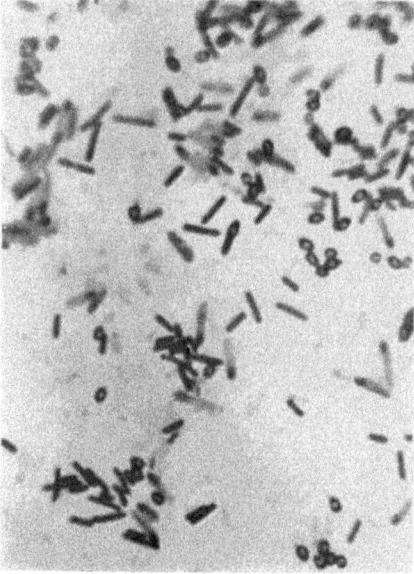

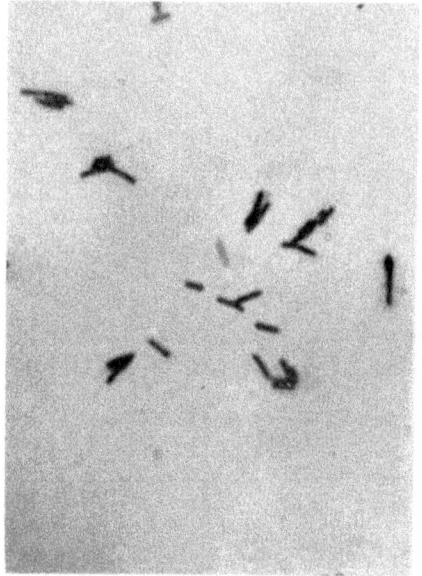

B. pantothenticus 14576

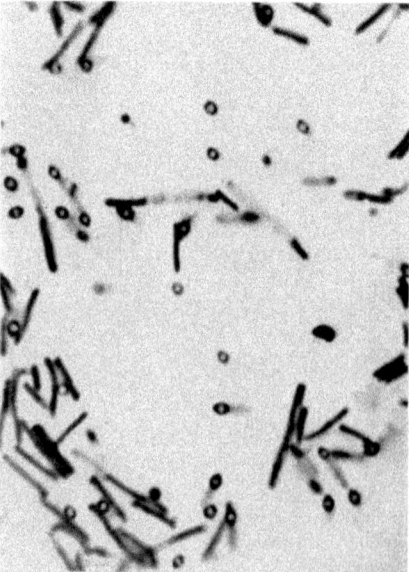

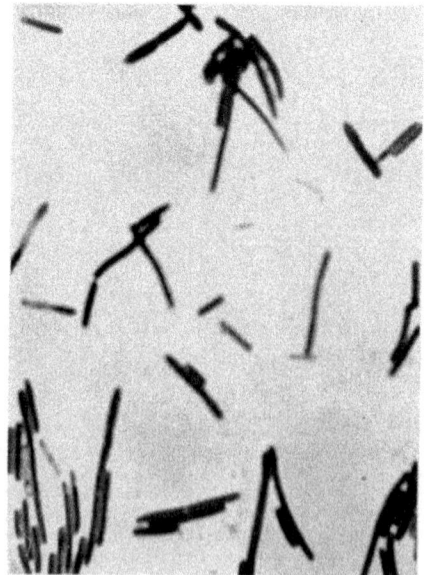

B. sphaericus 245

TABLE I

Numbers of strains of Bacillus species

Species	Number of strains	
	N.R. Smith's collection	American Type Culture Collection
B. megaterium	93	27
B. cereus	178	56
B. licheniformis	63	20
B. subtilis	201	46
B. pumilus	93	15
B. badius	2	1
B. coagulans	77	6
B. firmus	19	2
B. lentus	3	2
B. polymyxa	25	18
B. macerans	20	14
B. stearothermophilus	98	10
B. circulans	89	13
B. alvei	12	4
B. laterosporus	13	6
B. brevis	71	7
B. pantothenticus	6	1
B. sphaericus	46	10
B. pasteurii	4	4
Bacillus spp.	121	12
Total	1234	274

were sterilized at the end of the day, but the autoclave was not opened until the following morning, because if opened when the soil was still hot, the flasks frequently boiled over. The cool extract was carefully decanted and filtered through paper. Measured amounts of the filtrate (usually 300 ml) were autoclaved in small flasks for 20 min at 121° and allowed to stand at room temperature for 2 weeks or longer, during which time a sediment formed, leaving a clear supernatant extract to be decanted as needed.

Soil extract agar. The soil extract agar consisted of: 5 gm peptone, 3 gm beef extract, 15 gm Bacto-agar, 750 ml tap water, and 250 ml soil extract; pH was 7.0. The amount of soil extract used in this medium varied with the soil. When soil from a new source was extracted, small amounts of soil extract agar were prepared with 100, 75, 50, 25 (as above), or 10 per cent of soil extract. Each agar was inoculated with a culture of B. subtilis, and the percentage of soil extract yielding the best sporulation after 3 days of incubation was used in the preparation of this medium.

Nutrient agar. Nutrient agar was made up of 5 gm peptone, 3 gm beef extract, 15 gm Bacto-agar, and 1000 ml distilled water; pH was 7.0.

Preservation of cultures in sterile soil. Two parts by weight of air-dried soil and one part of air-dried humus were mixed and sifted through a coarse sieve. Nine parts of the mixture were added to one part by weight of $CaCO_3$ and poured into test tubes to a depth of approximately 2 cm. The tubes were plugged with cotton and placed in the autoclave in a slanting position to spread the soil in as thin a layer as possible. The soil was sterilized at 121° for one hour on three successive days. A sporulating culture was suspended in approximately 1 ml of sterile water; the suspension was carefully mixed with the sterile soil and smeared on the sides of the tube. When the soil was dry and no longer adhered to the sides of the tube, the tube was closed with a rubber stopper and stored at room or icebox temperature. The culture was revived by suspending a few particles of the soil in nutrient broth.

Microscopic examination. The smears of the cultures were air-dried and stained for 15 to 20 seconds with a solution of Hucker's (19) ammonium oxalate crystal violet diluted to one-half strength with distilled water. Unstained spores within and outside the sporangia were easily recognized, and a comparative estimation of their numbers was made (Table II).

Results

During the study of the classification of the aerobic, spore-forming bacilli, begun in 1936 by Smith, Gordon, and Clark (17) and terminated in 1951, a collection of 1234 strains (Table I) was assembled. Cultures of B. polymyxa and B. macerans were maintained on nutrient agar with 1 per cent potato starch; cultures of B. pasteurii, on nutrient agar plus 1 per cent urea; and those of the other species on nutrient agar. After sporulation, a set of cultures of each strain was stored in stoppered tubes at icebox temperature; another set in unstoppered tubes at room temperature; and a third set was preserved in sterile soil. The shredded agar used in the media during the project was rather crude and contained a great deal of sand, seaweed, and similar materials that had to be removed by filtering the agar through cotton or paper. During the time this agar was used, only eight strains were lost.

In 1950, we began preparation of new stock cultures of all the strains in the collection. Bacto-agar, purchased in 1949, was used in the nutrient, starch, and urea agars, but many of the cultures grew poorly and failed to sporulate. Because time and inclination did not permit determination of the substances in the crude agar that promoted spore formation by such a large number and variety of cultures, we compared sporulation by some of the cultures on nutrient agar made with Bacto-agar, nutrient agar made with

TABLE II

Sporulation by cultures of <u>Bacillus</u> species at 5 days

Strain number (N. R. Smith's collection)	Sporulation*		
	Nutrient agar	Nutrient agar with crude agar	Soil extract agar
B. megaterium			
239	++	++++	++++
343	-	'++	+
607	-	++++	++++
871	-	-	-
991	+++	++++	++++
B. cereus			
970, 973, 998	+++	++++	++++
974, 976, 996	++++	++++	++++
B. subtilis			
773	-	+	++
774	±	+++	++++
775	-	++	+++
968	+	++++	++++
971	-	++++	++++
972	-	++	++++
975	-	+++	+++
1088	-	-	-
B. pumilus			
840, 938	-	+++	+++
927, 980	++++	++++	++++
982	-	-	-
984	±	++	+++
985	±	+++	+++
990	-	++++	+++
997	±	++	++
B. coagulans			
T24	±	++	++
T27	+	+++	+++
T54	±	+++	+++
609	-	-	±
834	-	++	+
B. macerans			
373, 1098	+++	+++	+++
1099	+	+	+

TABLE II continued

Strain number (N. R. Smith's collection)	Nutrient agar	Nutrient agar with crude agar	Soil extract agar
B. stearothermophilus			
T37	+++	+++	++++
T59	++	++	+++
T63	t	++	++
T68	++	++	++
B. circulans			
832	+	+	++
842	-	-	-
924	+	++	+++
977	++++	++++	++++
1000	+++	+++	+++
B. brevis			
819	t	++	+++
889	-	-	-
925	++++	++++	++++
954	$\pm$	++	+++
B. sphaericus			
T156	+	++	++
592	-	+	++
966	++	++	++
967	+	+++	++
Bacillus spp.			
T7, T8	-	-	-
943, 965	t	++++	++++
964	$\pm$	+++	+++
986	+	+++	+++

* ++++ Predominance of sporangia or free spores in each field.
  +++ Spores formed by approximately one-half of rods per field.
   ++ Spores formed by approximately one-tenth of rods per field.
    + Approximately one spore per field.
    t Less than one spore per field.
    - Spores not observed in 15 to 20 fields.

crude agar, and nutrient agar made with Bacto-agar and 500 ml of soil extract per liter. Results of this study (Table II) show that, within most of the species, sporulation was subject to strain varia- tion. One of five cultures of B. megaterium, for example, did not form spores on any of the three media; one other culture sporu- lated abundantly or well on all three agars; and the remaining three cultures produced many more spores on agar made with crude agar or soil extract than on nutrient agar. On the whole, crude agar and soil extract effectively promoted spore formation by cultures of B. megaterium, B. subtilis, B. pumilus, B. coagu- lans, B. brevis, and B. sphaericus. With a few exceptions, cultures of B. cereus, B. stearothermophilus, B. macerans, and B. circu- lans sporulated equally well on the three media. The new stock cultures were then grown on soil extract agar (50 per cent soil extract) and preserved in sterile soil. When, after 6 years of storage, these were revived, all but five of the 1226 cultures were viable.

A recent examination of the American Type Culture Collec- tion's strains of Bacillus spp. (Table I) offered another opportunity to observe the effectiveness of soil extract on spore formation. Of the 274 cultures, 254 sporulated well or abundantly on soil extract agar after 3 to 5 days of incubation. Of the remaining 20 cultures, 9 formed fewer yet easily demonstrable spores. The cultures on soil extract agar that did not form observable spores were: B. megaterium 11478, 12138, and 13632; B. cereus 11949, 12826, and 12828; B. polymyxa 12712; B. macerans 7068; B. circulans 8241; B. laterosporus 4517; and Bacillus sp. 4978. Cultures of 11 strains did not sporulate after longer incubations.

Table III compares spore formation by some of the ATCC cultures on nutrient and on soil extract agars. We selected cul- tures at random of 18 species, grew the cultures in nutrient broth, and used the broth cultures as inoculum on the two agars. Approxi- mately 60 per cent of the 62 cultures formed more spores on soil extract agar than on nutrient agar. Soil extract effectively increas- ed sporulation by cultures of B. megaterium, B. licheniformis, B. subtilis (Pl. 1), B. pumilus, B. badius (Pl. 2), B. coagulans, B. firmus, B. lentus, B. circulans, B. laterosporus (Pl. 3), and B. pantothenticus (Pl. 4). Four cultures of B. cereus (Pl. 1) and two of B. polymyxa formed many more spores on soil extract agar than on nutrient; four other cultures of B. cereus and two of B. polymyxa sporulated well on both agars. On the whole, cultures of B. macerans, B. stearothermophilus, B. alvei (Pl. 2), B. brevis (Pl. 3), and B. sphaericus (Pl. 4) formed as many spores on nutrient agar as on soil extract agar. Variability among the cul- tures in their requirements for sporulation is apparent from a comparison of Tables II and III. Conclusions that might possibly

TABLE III

Sporulation by cultures of <u>Bacillus</u> species at 3 to 5 days

ATCC strain	Sporulation*	
	Nutrient agar	Soil extract agar
B. megaterium		
4531, 6458, 7703, 11561	-	++++
B. cereus		
2, 240, 13366, 13367	++++	++++
9620, 10206, 11986	+	++++
6604	-	++++
B, licheniformis		
8480	-	+++
9945, 11946, 13438	-	++++
B. subtilis		
6051, 6633, 6894	-	++++
7003	+	++++
B. pumilus		
71, 98	-	+++
72	-	++++
945	+	+++
B. badius		
14574	+	++++
B. coagulans		
7050, 8038	-	+++
10545	+	++++
11014	+	+++
B. firmus		
8247, 14575	-	+++
B. lentus		
10841	+	+++
B. polymyxa		
842	++	+++
7070	-	++++
8524	+++	++++
8525	-	+++
B. macerans		
843	+++	++++
7048, 8510, 8518	++++	++++

TABLE III continued

ATCC strain	Nutrient agar	Soil extract agar
B. stearothermophilus		
7953, 7954	++++	++++
8005	++	+++
10149	++++	+++
B. circulans		
61	-	+++
4513	++	++++
4516, 7049	+	++++
B. alvei		
6344	++++	++++
10871	++	+++
B. laterosporus		
64	+	+++
6456	++	++++
9141	+++	++++
B. brevis		
8185, 8186, 10068	++++	++++
10027	+++	++++
B. pantothenticus		
14576	-	+++
B. sphaericus		
245	+	+++
4525, 7063	+++	++++
7054	+++	+++

* See key, Table II.

be drawn from Table II--that cultures of B. cereus and B. circulans sporulate readily on nutrient agar and that cultures of B. brevis and B. sphaericus sporulate less readily on nutrient agar--are not confirmed in Table III.

Discussion

These observations on spore formation illustrate the fallacy of considering a small number of strains as representative of a genus or a species. The fact that sporulation by strains of one species does vary with media indicates that spore formation is dependent on several factors. The requirements for sporulation of one strain are not necessarily the requirements of other strains belonging to the same or other species.

These results on the usefulness of soil extract in promoting spore formation are similar to those reported by Smith and Worden (18) and by Lochhead (12) on the value of soil extract for growing soil bacteria. After a study of approximately twenty thousand isolates from various soils, Lochhead and his co-workers concluded that soil extract contains more than one factor essential to growth, and that each factor concerned can only be identified by the study of the bacteria from a large number of soils. We, too, are certain that the substances in soil extract that promote sporulation will eventually be identified and that the identification will require a study of many strains of each species. In the meantime, soil extract represents a valuable aid to microbiologists responsible for maintaining many strains of Bacillus spp. and more interested in keeping their strains than in using defined media.

The failure of some cultures of Bacillus spp. to sporulate on all media raises two questions for those of us interested in taxonomy and culture collections:

1. How many of the strains in our collections can form spores under different conditions of cultivation, and are misnamed? In two of the culture collections we have examined, we found sporulating cultures resembling B. subtilis that bore the labels of a goodly number of the species listed in Bergey's Manual of Determinative Bacteriology. Spores of B. subtilis are very liable to survive inadequate autoclaving and to overgrow cultures inoculated on improperly sterilized media. Among other misnamed cultures, we have encountered strains of lactobacilli that were B. coagulans and strains of Azotobacter spp. that were B. megaterium.

2. How many of our cultures are mixed with spore-forming bacilli? Unlike B. subtilis, which quickly supplants other bacteria, strains of some large-spore-bearing rods such as B. circulans grow in mixed cultures with no apparent effect on the growth and appearance of the second culture. Some strains of B. circulans and of related species produce a very thin film on soil media that can be seen only with difficulty by the unaided eye. Because they

are actively motile and form motile colonies that spread quickly over an agar surface, ordinary plating cannot separate them from colonies of more slowly growing bacteria. In our present collection of mycobacteria, nocardiae, and streptomycetes, the mycobacteria are the most liable to be contaminated by cultures resembling B. circulans.

To detect mislabelled cultures of Bacillus spp. and cultures contaminated with bacilli, we inoculate a slant of soil extract agar and after 5 to 7 days examine a Gram stain of the growth. The preparation and examination of a culture on soil extract agar is a part of our routine procedure of lyophilization and has provided a number of disagreeable surprises.

Summary

A collection of 1234 strains of Bacillus spp., assembled during a taxonomic study lasting 15 years, was maintained on media prepared with crude agar. During this period, only eight strains were lost. At the termination of the study, cultures of the remaining 1226 strains were grown on soil extract agar and preserved in sterile soil. After 6 years of storage, 1221 of the 1226 cultures were viable.

Of the American Type Culture Collection's 274 strains of Bacillus spp., 93 per cent sporulated well or abundantly on soil extract agar.

In a comparison of spore formation on soil extract and nutrient agars of 121 cultures representing 18 species of Bacillus, soil extract effectively increased sporulation by 54.5 per cent.

This study was supported in part by the National Science Foundation and the American Type Culture Collection. We greatly appreciate this assistance and also acknowledge our indebtedness to Dr. N. R. Smith who proposed soil extract for promoting spore formation.

References

1. Brady, R. J., Chan, E. C. and Pelczar, M. Jr.   Sporulation of Bacillus subtilis grown in association with Erwinia atroseptica. J. Bacteriol. 81: 725-729. 1961.

2. Campbell, L. L. Jr. and Sniff, E. E.   Effect of subtilin and nicin on the spores of Bacillus coagulans. J. Bacteriol. 77: 766-770. 1959.

3. Charney, J. C., Fischer, W. P. and Hegarty, C. P.   Manganese as an essential element for sporulation in the genus Bacillus. J. Bacteriol. 62: 145-148. 1951.

4. Curran, H. R.   The mineral requirements for sporulation. Publ. Amer. Inst. Biol. Sci. No. 5, pp. 1-9. 1957.

5. Fields, M. J.   The effect of Oidium lactis on the sporulation of Bacillus coagulans in tomato juice. Appl. Microbiol. 10: 70-73. 1961.

6. Gollakota, K. G. and Halverson, H. O. Biochemical changes occurring during sporulation of <u>Bacillus cereus</u>. I. Inhibition of sporulation by a picolinic acid. J. Bacteriol. 79: 1-8. 1960.

7. Ierusalimski, N. D. and Egorova, L. A. The relationship of <u>Bacillus megatherium</u> to the condition of its life cycle. Mikrobiologiya (trans.) 29: 241-244. 1960.

8. Kanai, M. Studies on the influence of calcium ion upon the sporulation of <u>Bacillus subtilis</u>. Bull. Tokyo Med. Dental Univ. 6: 105-120. 1959.

9. Kolodzie, B. J. and Slepecky, R. A. A copper requirement for the sporulation of <u>Bacillus megaterium</u>. Bacteriol. Proc. p. 48. 1962.

10. Knaysi, G. A study of some environmental factors which control endospore formation by a strain of <u>Bacillus mycoides</u>. J. Bacteriol. 49: 473-493. 1945.

11. Knaysi, G. The endospore of bacteria. Bacteriol. Rev. 12: 17-77. 1948.

12. Lochhead, A. G. The nutritional classification of soil bacteria. Proc. Soc. Appl. Bacteriol. 15: 15-20. 1952.

13. Long, S. K. and Williams, O. B. Factors affecting growth and spore formation of <u>Bacillus stearothermophilus</u>. J. Bacteriol. 79: 625-628. 1960.

14. Ordal, J. Z. The effect of nutritional and environmental conditions on sporulation. Publ. Amer. Inst. Biol. Sci. No. 5, pp. 18-26. 1957.

15. Schmidt, C. F. Spore formation by thermophilic flat sour organisms. I. The effect of nutrient concentrations and the presence of salts. J. Bacteriol. 60: 205-212. 1950.

16. Shinagawa, T. Studies on the sporulation of <u>Bacillus cereus</u> var. <u>mycoides</u>. J. Osaka City Med. Center 9: 4047-4054. 1960.

17. Smith, N. R., Gordon, R. E. and Clark, F. E. Aerobic spore-forming bacteria. U. S. Dept. of Agr. Monograph No. 16. 1952.

18. Smith, N. R. and Worden, S. Plate counts of soil organisms. J. Agr. Research 31: 501-517. 1925.

19. Society of American Bacteriologists, Committee on Bacteriological Technic. Manual of Microbiological Methods, p. 13. McGraw-Hill Book Company Inc., New York. 1957.

20. Tomscik, J., Bouille, M. and Baumann-Grace, J. B. Reaction specifique de l'exosporium chez <u>Bacillus cereus</u> and <u>Bacillus anthracis</u>. Schweiz. Z. allgem. Pathol. u. Bakteriol. 22: 630-640. 1959.

DISCUSSION I

by H.P.R. Seeliger
Hygiene Institute
Bonn, Germany

Opening the discussion on the topic "Methods for the Preservation of Bacteria and Actinomycetes" I want to congratulate first the authors for the interesting contribution to which we have been listening. It is this type of basic work that needs to be done if we want to understand better the principles of the preservation of bacteria.

Although lyophilization has brought great progress to the maintenance of bacterial and fungal stock culture collections, it has by no means solved all the problems. Before lyophilization is done, one must be sure that the cultures possess all their typical morphological, biochemical, serological and biological characters and that they are really pure. The majority of bacteria of medical importance fortunately remain viable on simple nutrient media for long periods of time and I wholeheartedly subscribe to the statement made in this respect by Prof. Murray in the previous session.

Sporulation is but one important criterion that characterizes certain bacterial groups, and many of us have experienced that this property is not rarely lost on artificial media and that spore formation depends on a variety of more or less unknown factors. If spores are produced, preservation of strains offers no difficulties, for spores remain viable in the dry state over long periods of time, for instance in sterile sand mixed with blood. From the Hygiene-Institute of Gottingen, Germany, it was recently reported that spores of Bacillus anthracis and B. mesentericus dried on silk threads were viable after 65 to 70 years preservation (Brandis, H.: Zbl. Bakt. I. Orig., 177, 434, 1960) and at Berlin, spores of the anthrax bacillus preserved by Robert Koch, have germinated and produced typical cultures. You will recall that Dr. Sneath mentioned yesterday that spores have survived in soil for 300 years.

Referring to the use of soil extracts as stimulating agents for spore production I should like to draw your attention to similar experiments of mycologists with fungus cultures. Many pathogenic fungi have a tendency, on artificial media, to convert from their typical and characteristic form into a pleomorphic stage. Thereby they often lose their ability to produce bodies of mass reproduction, i.e. micro- and macroconidia. Transfer of such cultures to organic materials such as sterilized feathers or soil, has proved a valuable aid in maintaining or restoring typical morphology including formation of reproductive cells of increased resistance to untoward environmental conditions. Although I have not used the

word "spores" for such micro- and macroconidia or ascospores of
fungi, in order to avoid confusion with bacterial spores, one might
ask whether the formation of such structures in bacteria and fungi
may not be based on similar nutritional requirements, varying with
species and even from strain to strain.

Dr. Gordon has apparently not encountered major difficulties
in sterilizing soil extracts by autoclaving them for one hour at
121°C. But will this procedure always prove to be adequate for the
killing of thermoresistant spores? Some years ago in reports from
Graz, Austria and Berlin (Kurzweil, H.: Arch. Hyg., 138, 77, 1954;
Gelinsky, E. and G. Lockemann: Zbl. Bakt. I. Orig., 156, 305, 1951)
the relatively frequent occurrence of highly thermoresistant spores
in certain fertilized soils was emphasized. Could then the occasional presence of such spores represent a serious handicap for
the use of soil agar and soil extracts? Instead of filtering through
paper, one might perhaps prefer to sterilize the supernatants of
boiled extracts by Seitz filtration. Or do such filters remove or
diminish substances essential for inducing spore formation?

Some of the statements made by the speaker should be stressed again because they are of general validity. Always a rather
large number of different strains of one species should be investigated before any relevant conclusions can possibly be drawn as to
the use of the medium of choice.

The presence of spore-formers, such as B. subtilis and B.
circulans, in cultures in collections is widely underestimated. This
may be due to primary contamination or to secondary infestation
as a consequence of unsatisfactory media sterilization procedures
and last, but not least, by misnaming cultures due to lack of spore
formation on artificial media.

Such observations are not at all restricted to spore producing bacteria. The anaerobic and microaerophilic Actinomyces
strains such as A. israelii and related species, are very frequently
mixed with and finally outgrown by other anaerobic bacteria with
which they occur associated under natural conditions. Association
of A. israelii with Corynebacterium acnes has apparently been the
cause of keeping C. acnes strains in several culture collections
under the designation of A. israelii. Although by means of lyophilization the anaerobic Actinomyces can now be easily preserved,
their initial purification offers sometimes great difficulties even
to experienced workers in this field.

When a few years ago Hackenthal and Bierkowski (Zbl. Bakt.
I. Orig., 162, 160, 1955) drew attention to their observations that
many streptococcus, staphylococcus and other cultures deposited
in recognized collections, consisted of mixtures in various propor-
tions of related species or biotypes, these reports were first received with due caution. In the meantime, however, additional

evidence has been obtained that quite a number of reference strains
may contain very small admixtures of other organisms. These may
remain undetected for long periods of time. Observations made
during the past years indicate that enterobacteriaceae cultures are
frequently not pure. From one allegedly pure salmonella culture
5 different serological types could be separated. The presence of
at least 2 sero-types in so-called pure cultures of salmonella is,
according to our own experience, surprisingly frequent.

Some of these observations can be explained by insufficient
procedures of strain purification. The frequency and degree of
contamination of cultures with other organisms depends largely on
their origin. This may be illustrated by one example: E. coli is
generally, but erroneously, considered as the predominant organism in faecal material, although in most stools the proportion of
E. coli to other cultivable bacteria such as other enterobacteriaceae
and predominantly strict anaerobes, may be as low as 1:10,000.
Invariably viable organisms of other species will be present in E.
coli strains isolated from the primary plates. Some of these
anaerobic bacteria seem to live in some sort of symbiosis with E.
coli and related organisms, and may be carried through many passages. Use of selective, i.e. inhibitory media, and transfer from
solid to liquid media with rapid subculturing is of great help in the
purification process.

Extending the final remarks of Dr. Gordon on the frequent
contamination of mycobacteria, one also might mention the great
difficulties that arise for the mycologist in biochemical studies
when fungous cultures are intimately mixed with spore-forming and
other bacteria.

DISCUSSION II

by Mortimer P. Starr
Department of Bacteriology
University of California
Davis, U.S.A.

Albeit there is room for concern about the paucity of depend-
able data on the rationale of some techniques, the contributions of
previous speakers at this conference show that several procedures
are available and in use for the preservation of ordinary bacteria.
The bacteria which cause diseases of plants are ordinary bacteria:
pseudomonads, enterobacteria, rhizobia, coryneforms, bacilli, and
actinomycetes (4). Indeed, apart from plant pathogenicity, there is
not much to distinguish these creatures from their non-virulent
counterparts. Therefore, most of the problems which are here

termed "peculiar", stem from the one unique property of being virulent to plants and, in turn, from an incredible overweighting in nomenclatural usage of one ecological fact, namely, that they are isolated from diseased plants.

The two major specialized phytobacterial collections, the National Collection of Plant Pathogenic Bacteria (NCPPB), Rothamsted, England and the International Collection of Phyto-pathogenic Bacteria (ICPB), Davis, California, both use lyophiliza-tion. Largely because of the excellent record of revivability during our quarter-century's experience with lyophilization of phytopatho-genic bacteria, we are very satisfied--indeed, to the point of com-placency--with this method of preservation. Some concern has been expressed by correspondents on the theme that there may be --incident to the freezing-and-drying procedure--some mutagenesis or selection in the direction away from plant pathogenicity. It has not been possible, however, to uncover authenticated instances to substantiate this notion, though my file relative to this conference contains a number of experimentally unsupported statements to that effect from reputable workers. This point has apparently not been studied systematically; however, incidental to our own diverse investigations (4), we have not observed alterations in phytopatho-logical, biochemical or metabolic properties directly attributable to lyophilization; a direct evaluation (3) shows that freezing itself has no mutagenic effect on another genetic trait in Xanthomonas. To avoid the presumed danger of selection by lyophilization--or simply to be different--the bacterial phytopathologists have re-ported an astonishing array of conservation techniques (cf. Ref. 2 for an excellent summary). Some of these clearly succeed in the avowed purpose of maintaining plant virulence at whatever level and type the culture possessed originally; others seem to fail very badly, either in the experience of the individuals who suggested the method, or in the experience of others. These procedures range the gamut: maintenance on various artificial culture media which provide anything from a starvation to a gourmet diet; holding on natural materials, such as sterilized soil, portions of plants, entire live plants, and assorted other substrates in that genre; storage at low temperature, including freezing in glycerol (3); exposure to the air, or protection from the atmosphere by a layer of mineral oil or by other means; or, mirabile dictu, suspension in distilled water at ambient temperatures. This latter procedure is reported by reput-able scientists to maintain for years not only the viability but also the virulence of plant bacteria, and involves nothing more compli-cated than the placement of cells in screw-cap vials of sterilized distilled water. Hence, we cannot overlook the fact that conserva-tion procedures for bacterial phytopathogens do, in fact, extend from thorough desiccation to thorough drowning!

The most serious problems in maintaining a collection of
phytopathogenic bacteria, however, lie in areas away from preservation; namely, in acquisition and distribution. The curators of
NCPPB and ICPB join in the generalized curatorial lament about
the difficulty of acquiring ancient and contemporary nomenclatural
types, neotypes, and other representative cultures. Even rather
heroic procurement efforts have resulted in failure, despite the
unambiguous Recommendation 9b of the International Code of
Nomenclature of Bacteria and Viruses. The suggestion has been
made that new taxa be considered validly published only when
accompanied by deposition of type cultures in recognized active
culture collections. It is hoped that this recommendation will be
converted into a rigorously enforced nomenclatural regulation.

At this point, I must respectfully disagree with the curatorial
philosophy expressed by at least one colleague at this conference
on the theme of what to acquire. In the present state of ignorance
on the speciation of phytopathogenic bacteria, it is obligatory to
acquire everything! Whether one compares by classical methods
or by computers, many of the so-called "species" of plant disease
bacteria are clearly either phytopathogenic formae speciales or
physiological varieties. In accordance with the phytopathological
doctrine which has whimsically been entitled "the new host--new
species concept" (4), several dozen "new species" are described
in a typical year. That is to say, somebody describes a phytopathogenic bacterium occurring on a plant species, which had not
previously been reported as the victim of a bacterial disease, and
gives it a "new" species name following only cursory comparison
with already named species. Acquisition of these cultures, whether
they are alleged to be new or old species or sub-species, is essential to the ultimate development of a rational taxonomy. I most
emphatically would not presume to weight the ultimate analysis by
exerting the suggested curatorial prerogative of excluding cultures.
Naturally, this means extra labour and perhaps uncritical storage
--but these are nonetheless made necessary by the very nature of
the material with which plant bacteriologists work.

One additional problem peculiar to plant pathogens also
stems from their virulence and should be mentioned before an
international group of this sort. The distribution of these cultures
is technically and legally under regulations administered by
assorted international, national (1), and local quarantine agencies.
Licenses are required for each movement of phytopathogenic
microbes. Such licenses are generally issued to reputable workers
by the competent agency. It means simply filling out forms and
waiting--naturally, filling out forms is repugnant to most acade-
micians, and waiting is anathema to any scientist. Nonetheless,
the law is the law. An impatient worker or one ignorant of these

rules might break the law by moving cultures without licenses. In actual practice, there is great variation in the stringency with which the quarantine regulations are enforced or obeyed, the consequent dual standards do nothing to achieve a sane approach to regulatory practices. Much better, would be the development and rigorous enforcement of acceptable regulations which serve both to protect the public health and to further the quite laudable aims of responsible plant bacteriologists.

References

1. Congress of the United States of America. Federal plant pest act. Public Law 85-36, 85th Congress, S. 1442, May 23, 1957.
2. Kelman, A. The bacterial wilt caused by Pseudomonas solanacearum. North Carolina Agricultural Experiment Station. Tech. Bull. 99: 1-194. 1953.
3. Quadling, C. Preservation of Xanthomonas by freezing in glycerol broth. Canad. J. of Microbiology 6: 475-477. 1960.
4. Starr, M.P. Bacteria as plant pathogens. Annual Review of Microbiology. 13: 211-238. 1959.

GENERAL DISCUSSION

Miss Gordon, USA - To elaborate on a point that Dr. Seeliger raised regarding the sterilization of soil extracts. Our soil extract is autoclaved three times before it touches the culture - the soil and water is autoclaved for one hour, then the filtrate is autoclaved for 20 minutes, stored and again autoclaved after its addition to the agar.

K.S. Zinneman, England - One of the most difficult things to do, in the clostridia group, is to induce spore-formation in Cl. welchii. Dr. Gordon, do you have any experience with that organism in soil extract?

Miss Gordon, USA - I do not have any information on that topic.

H. Katznelson, Canada - Dr. Gordon, have you observed any differences in the nature of the soil extracts with regard to the germination of spores? The reason I ask is that, in studies on the germination of some of these species, we found that certain soil extracts would not permit germination but the fortification of these extracts with casamino acids, for example, would permit it.

Miss Gordon, USA - I haven't any information on the germination of spores. However, I do know that different lots of soil extract can give different amounts of sporulation.

K.S. Zinneman, England - We have occasionally come across contaminated cultures from culture collections, as Dr. Seeliger mentioned. I wonder what Dr. Seeliger would think of culture collections using single cell isolation from cultures.

H.P.R. Seeliger, Germany - I had a remark in my discussion mentioning single cell isolation but I cut it out after some discussions I had yesterday. You will recall that the question was dealt with yesterday and that a warning was issued that by the use of single cell isolates, we might lose important variants. However, I don't see any obstacle in using single cell isolates if these are thoroughly investigated so that we are sure that they contain all the original properties we ascribe to a certain strain.

S.T. Cowan, England - I think that single cell isolation is a thoroughly dangerous procedure - I don't even recommend single colony isolation. Thus, our procedure is to take a sweep of several colonies.

The thing that you have to guard against is what I call cryptic contamination - the sort of contaminants that don't appear within 24 hours when you grow the culture say at 37°C. Our procedure now is to leave all plates for a week at room temperature and then examine them.

H.P.R. Seeliger, Germany - As far as Dr. Cowan's recommendations of strain purification are concerned, I would like to add that one also should think of using anaerobic methods. We did a few experiments along this line - diluting standard strains of E. coli up to say 10^6 or 10^8 and then culturing them in Brewer's anaerobic jars on blood agar for 5 to 7 days. We were amazed how often we found gram-negative, anaerobic, non-spore-forming organisms.

P.A. Hansen, USA - There is one source of contamination which I don't think is always appreciated. In one-step lyophilization there can be considerable carry-over by a blast of the suction from one vial to another. Thus, if you are lyophilizing several strains in the same batch, this may act as a source of danger unless you have some provision for cotton plugs to take care of the situation.

CULTURE COLLECTIONS OF ALGAE

by R. C. Starr
Indiana University
Bloomington, U.S.A.

The algae include a vast assemblage of organisms in whose great diversity of form and function lies a tremendous research potential. For example, the plant bodies of algae may be single cells or multicellular structures over 300 feet in length, or any of a variety of sizes and cell arrangements; nutrition in the algae varies from obligate autotrophy to obligate heterotrophy, from saprophytism to parasitism; genetic systems can be studied in life cycles involving either a haploid organism or a diploid organism, or an alternation of both types in the same cycle. The research potential remains largely unexplored and will remain so until algal cultures become more readily available. Researchers interested in particular systems, such as genetics, already find themselves severely limited by the few organisms on hand. Such individuals more often are not trained in phycology and/or the techniques of algal culture, and so they must rely completely on others to point out suitable organisms, isolate them, and learn to control them in culture.

Existing Facilities

Although a number of private laboratories maintain stocks of algae as part of their research programs, only a few collections have been established to serve as centers of distribution to the general scientific community. These include the following:

Charles University, Prague	Dir.: S. Prat
Cambridge University, England	Dir.: E. A. George
Göttingen University, Germany	Dir.: E. G. Pringsheim
Laboratoire de Cryptogamie, Paris	Dir.: P. Bourrelly
Indiana University, Bloomington	Dir.: R. C. Starr

Of the above collections only those at Göttingen and Bloomington are continuing to accept cultures from investigators. The lack of collections and the limitation of the existing ones are due in part to the difficulty of obtaining support for such projects. For example, the lack of funds has as yet prevented the establishment of a collection in Japan. Until granting agencies become more aware of the need for extensive collections or until the economic importance of the algae is developed so that industry will be willing to support such collections, the picture may not be expected to change radically.

Current Methods of Preservation

There have been embarrassingly few efforts to improve methods of preservation in the algae. The current method employed

in the collections at Cambridge, Göttingen, and Bloomington involves lowering the temperature, decreasing the illumination, and
preventing excessive water loss. At Bloomington the temperature
is lowered to 10 C, the light is reduced to less than 50 ft.-c. intensity for a 6 hr light: 18 hr dark cycle, and screw cap tubes are
employed to prevent excessive evaporation. Similar devices are
used at the other collections. Many bacteria-free cultures can be
maintained for periods up to and exceeding 12 months without
transfer. The extremes require continued cultivation at 20 C,
under illumination of 250 ft.-c. intensity in unsealed containers,
and a transfer frequency of 10 days to 2 weeks.

With an increasing interest in algae for research there is
every reason to believe that culture collections will be called upon
to handle such large numbers of strains that the current methods
will not be feasible even if they were considered the most reliable
for maintaining the genetic purity of a strain which, of course, they
are not. There have been relatively few efforts made to test with
large numbers of species other methods of preservation which
have proved so successful with other groups of organisms.

Daily and McGuire (1) reported on the lyophilization of 32
different algae belonging to the Chlorophyta, the Chrysophyta, and
the Cyanophyta. Horse serum was used as the suspending agent
and the procedures followed were those used routinely at the Lilly
Research Laboratories for the preservation of other types of
micro-organisms. Of the 32 cultures, 24 (75%) were viable when
tested within 24 hours following lyophilization. Duplicate tubes
when checked from 4 to 10 months after lyophilization also proved
to be viable.

Daily and McGuire checked the percentage of cells of three
species remaining viable after the desiccation and found the
following:

Scenedesmus obliquus .... 0.025% viable cells
Bracteacoccus cinnabarinus ...... 3.166% viable cells
Chlamydomonas pseudococcum ... 0.013% viable cells

They point out that these figures are similar to those obtained
with some bacteria (2).

Watanabe (4) lyophilized cultures of the blue-green algae
Tolypothrix tenuis and Calothrix brevissima and the green alga
Chlorella ellipsoidea. The Chlorella failed to grow immediately
after the treatment; the results with the blue-green algae are
given in Table I. The nitrogen-fixing capacity of the blue-green
algae was found to have remained unimpaired by the lyophilization
process.

It has long been known that air-dried soil samples upon inoculation into proper media under suitable environmental conditions
will yield algal populations even after 5 years or longer.

TABLE I

Change of viability of blue-green algae during
preservation under lyophilization

Protective substance	Human serum albumin	Bovine serum albumin
	Tolypothrix tenuis	
Time after lyophilization (months)		
0	92*	32
3	79	26
18	71	8
24	58	0
	Calothrix brevissima	
0	96	58
3	87	42
18	46	17
24	42	0

*refers to the ratio of viable cells in the lyophilized
sample to that in fresh sample.

Watanabe (4) was successful in adapting a modification of this
natural process to the preservation of mass cultures of Tolypothrix
tenuis and Calothrix brevissima grown for seeding rice fields. The
algae were inoculated onto a special volcanic earth known commercially as "Kanumatsuchi" and used for "Bonsai" culturing. Prior
to inoculation the earth was washed with distilled water, soaked
with nitrogen-free medium, and sterilized. The earth was then
mixed with equal amounts of a concentrated algal suspension, placed
in either tubes of polyvinyl sheeting or Petri dishes, and illuminated
with weak light. After 4 week's growth the algal-covered soil particles were transferred to vinyl bags or to cotton-plugged bottles
and stored at room temperature. The results are given in Table II.

TABLE II

Change in growth capacity and water content of blue-green algae
during preservation on surface of fine porous gravel

Time after inocul. (years)	*Tolypothrix tenuis* Growth	Water content (per cent)	*Calothrix brevissima* Growth	Water content (per cent)
0	+++	62. 9	++	61. 3
1	+++	59. 2	++	52. 6
2	++	14. 3	++	14. 9
3	++	11 5	+	13. 2
4	+	10. 0	+	12. 1
5	+	9. 8	t	10. 2

Venkataraman (3) attempted with some success the use of sand seeded with algae and then sun-dried. Although the blue-green alga Nostoc sp. proved to be viable after two years, Chlorella sp., Selenastrum westii, and Scenedesmus sp. failed to grow immediately after the treatment.

W. A. Clark (personal communication) of the American Type Culture Collection has found that algae can be successfully preserved using the low temperatures of liquid nitrogen, but a definitive investigation of this method has not been made.

The Future

The service which one may expect from culture collections of algae in the future will depend in great part on our efforts and successes in two areas. First, it is imperative that some earnest effort be made to investigate all the possibilities of long term preservation of the algae. Second, in order to have worthwhile strains in the collections there must be more research on the cultivation of algae of all kinds. An increased knowledge of the nutrition, the reproductive capacity and its control, the dormancy of zygospores and other resistant cells, the temperature requirements, the light effects other than photosynthesis, the extracellular products, and the mechanism and control of morphogenesis will be necessary to insure a good supply of algal cultures suitable for research in many disciplines where biological materials with desired qualities are yet lacking.

References

1. Daily, W. A. and McGuire, J.M. Preservation of some algal cultures by lyophilization. Butler University Botanical Studies XI: 139-143. 1954.
2. Hornibrook, J. W. A simple, inexpensive apparatus for the desiccation of bacteria and other substances. Jour. Lab. Clin. Med. 34: 1315-1320. 1949.
3. Venkataraman, G. S. A method of preserving blue-green algae for seeding purposes. J. Gen. Appl. Microbiol. 7: 96-99. 1961.
4. Watanabe, A. Some devices for preserving blue-green algae in viable state. J. Gen. Appl. Microbiol. 5: 153-157. 1959.

DISCUSSION I

by E. A. George
Cambridge University
Cambridge, England

Algae

Professor Starr has covered most of this rather sparse field, leaving me rather little to say.

However I would like to add to the record that in 1948 Professor Kluyver sent Professor Pringsheim lyophilised cultures of Chlorella, Chlorococcum, and Trebouxia. In Cambridge, Annear and George (unpublished) confirmed the work of Daily and McGuire, once again using a method of vacuum drying developed for bacteria. We were unable to extend the method to larger or more complex organisms such as Euglena and Volvox. I agree with Professor Starr that there is an urgent need for further investigation in this very promising field. I hope to be able to try deep-freezing methods very soon.

Free-living Protozoa

As far as I am aware no free-living protozoa are kept other than as active cultures maintained by methods similar to those mentioned by Starr for algae. We have about 60 strains in the Cambridge Collection - a trivial number compared with our thousand algal strains but culture for culture they are often much more laborious to maintain. Many laboratories successfully keep various protozoa for teaching purposes in very crude cultures. Attempts to improve these cultures e.g. to make them monoxenic are often failures. Once again there is a crying need for research - though with more resources many more species could doubtless be kept by present methods.

Entozoic Protozoa

Owing presumably to their medical importance more progress has been achieved in the preservation of these than other groups of protozoa or algae. This is all the more creditable to the workers concerned as in general these creatures are more difficult to culture than the free living forms.

The most promising method seems to be freezing in a suitable medium, such as dilute glycerol at -79°C and storing at this temperature. Among the notable successes in this field are the preservation of Entamoeba histolytica by Fulton and Smith (1) and of Trypanosoma spp. by Polge and Soltys (2). Annear (3) on the other hand preserved Strigomonas oncopelti successfully using a vacuum drying method.

As an outsider in the field of parasitology I am struck by the lack of a general collection of these protozoa. At present they are maintained and made available by a number of laboratories, academic, medical and commercial; but it seems hardly fair to expect such laboratories to assume the responsibility of permanently maintaining strains which may be no longer of interest to themselves.

Bryophyta and higher plants

Many bacteria-free strains of mosses and liverworts are maintained both at Prague and Cambridge. They are treated exactly as the algae. A few ferns and flowering plants have been grown axenically - often more for fun than anything else. At the moment we have only <u>Wolffia</u> <u>arrhiza</u> in the Cambridge collection.

References

1. Fulton, J. D. and Smith, A. U. Ann. Trop. Med. Parasit., 47: 240. 1953.
2. Polge, C. and Soltys, M. A. Trans. Roy. Soc. Trop. Med. and Hyg. 51:519. 1957.
3. Annear, D. I. Nature 178:413. 1956.

DISCUSSION II

by Norman D. Levine
University of Illinois
Urbana, U. S. A.

Relatively little research has been done on the long-time preservation of protozoa, partly because they present some special problems. They can very rarely be lyophilized successfully, and simple freezing--either rapid or slow--kills the great majority of species. Hence most of the work has been done with glycerol or some similarly acting adjuvant such as ethylene glycol or, recently, dimethyl sulfoxide.

Because of the difficulty of preserving most protozoa and because of the necessity of transferring their cultures frequently, there are no extensive culture collections of them (except for the phytoflagellates) such as those at Göttingen, Indiana University and Cambridge University. Individual investigators, however, maintain small, specialized collections. Quite recently, a committee of the Society of Protozoologists under the chairmanship of Dr. Luigi Provasoli prepared a catalog of cultures of protozoa maintained by various investigators in many parts of the world (1). It not only lists the names of available species and strains of free-living and parasitic protozoa, but it also gives their date and place of isolation, the names of their isolators, and quite extensive notes on culture media, frequency of transfer, etc.

Not all species can be preserved by freezing with glycerol. Flagellates appear relatively easy to preserve, and some amoebae, <u>Toxoplasma</u>, <u>Plasmodium</u> and a few other protozoa have been preserved; but attempts to freeze ciliates appear so far to have failed.

Survival is affected by many factors. We have studied some of these quantitatively, using the flagellate, <u>Tritrichomonas</u> <u>foetus</u>. While the same conditions may not prevail for other protozoa, our findings may serve as a guide for work on them. Most of our work has been done in solutions containing 1.0 M glycerol, although some other adjuvants have been used also.

    In parallel tests at a concentration of 1.0 M in CPIM (cysteine-peptone-liver infusion-maltose-serum) medium, following slow freezing to -20 C and storage at that temperature for 1 day, ethylene glycol was 80-90% as effective as glycerol, 1,2,3,4-butanetetrol was 67% as effective, 1,2-propanediol was 57%, 2,3-butanediol was 14%, mannose was 6% and mannitol was 1% as effective as glycerol; glycerol monoacetate and polyvinyl alcohol were ineffective, and 2,3-dimercaptopropanol was toxic (7).

    _T_. _foetus_ cannot withstand rapid "snap" freezing to either -20 or -76 C even in the presence of glycerol. Slow freezing to either temperature at the rate of about 1 degree C per minute gives good results (7).

    While _T_. _foetus_ can be grown indefinitely at 37 C in the presence of 10% glycerol and while the same concentration of glycerol is protective when the protozoa are frozen, survival is extremely poor if the protozoa are held for a few hours with the glycerol in the refrigerator before they are frozen. There is thus a critical zone near 4 C in which glycerol appears to be toxic and through which the protozoa must pass fairly rapidly for success (4).

    A glycerol concentration of 1.0-1.1 M (9-10%) gives best results when freezing in CPIM medium; 0.55 M (5%) is about as effective, but 0.28 M and 0.14 M are much less effective (7).

    If the sodium chloride concentration of the medium is increased, protozoan survival upon freezing is decreased, but increasing the glycerol concentration up to a maximum of 1.5 M will then increase the survival of the protozoa; above this level, the glycerol is toxic (9). This reciprocal relationship between sodium chloride and glycerol concentrations is partially in accordance with the "salt buffering" theory of glycerol's mode of action in protecting cells against freezing injury.

    Glycerol does not protect _T_. _foetus_ against death if the protozoa are frozen during the first half of their growth period. In experiments in which the population peak occurred an average of 28.1 hours after inoculation, the average culture age at which the first protozoa survived freezing was 20.3 hours, at which time the population had reached 53% of its peak. The optimum culture age for survival after 7 to 15 days of frozen storage at -21 C averaged 32.3 hours, at which time the protozoan population had decreased to 82.7% of its peak (8).

    The duration and temperature of equilibration with glycerol before freezing affect subsequent survival. When equilibration is carried out at room temperature, survival upon subsequent freezing is better following rapid equilibration (glycerol added all at once; equilibration time, 1 hour) than following slow equilibration

(1/6 of the final amount of glycerol added each hour for 6 hours; equilibration time, 7 hours). Survival is poor following either rapid or slow equilibration at 4 C (4).

Addition of 0.5% vegetable lecithin to the medium in the hope of decreasing the brittleness of the cell membrane did not increase the survival rate of T. foetus. This type of lecithin was harmless to the protozoa, but 0.5% egg lecithin or 0.1% animal lecithin killed them in cultures at 37 C (9). Buffering the storage medium to pH's 6.3 to 7.1 with glycylglycine increases survival upon freezing, but buffering to the same pH's with triethanolamine has no significant effect upon survival, and buffering to pH's 7.1 to 7.5 with phosphates decreases survival upon freezing (4).

Other constituents of the suspending medium also affect survival of the protozoa. T. foetus will not survive freezing in the presence of glycerol after it has been grown in the beef extract-glucose-peptone-serum medium of Fitzgerald, Hammond and Shupe (3), but it survives well after it has been grown in Johnson and Trussell's (5) CPIM (cysteine-peptone-liver infusion-maltose-serum) medium or in Diamond's (2) trypticase-yeast extract-maltose-cysteine-ascorbic acid-serum medium without phosphates. Survival is much better in this last medium after the protozoa have grown in it than it is in fresh Diamond's medium or in physiological salt solution; there is no significant difference in survival between the latter two suspending media. Presumably some product or products of the trichomonads' metabolism have an additional protective action which supplements that of glycerol (6).

T. foetus survives about as well when frozen and stored at a nominal temperature of -20 as at one of -79 C (7). The type of freezer employed and the temperature it actually maintains have a marked effect on survival. Survival upon storage was better in a chest-type freezer in which a standard recording thermometer showed that the temperature remained constant at -21 C than in an upright freezer in which cyclic temperature fluctuations between -23 and -25 C were recorded or in a dry-ice chest with a nominal temperature of -72 C. By use of special thermocouples inside the freezing tubes, it was found that the temperature actually fluctuated between -22 and -24 in the chest-type freezer; that it fluctuated between -19 and -30 in the upright freezer, but rose as high as -2 when the door was opened and the samples were removed; and that it fluctuated markedly in the dry-ice chest as the dry ice melted and was replaced, sometimes rising as high as -27 C. The poorer survival in the latter two freezers was considered due to temperature fluctuation. The protozoa die off slowly during storage at the above temperatures (4).

One explanation of the slow "storage death" of cells and protozoa is that it is due to gradual denaturation of the proteins (10).

Continued slow metabolism is probably also involved. The anti-metabolites malonic acid, sodium fluoride and sodium iodoacetate in concentrations too low to affect the protozoa at 37 C decrease the survival rate of the protozoa upon storage at -21 C; upon freezing, their concentrations in the liquid phase of the suspending medium increase enough to affect metabolism (9).

Both protein denaturation and slow metabolism are temperature dependent, and their rates decrease exponentially with decreasing temperature. They become negligible at temperatures around -80 C and presumably reach zero at the glassy transformation temperature of ice (approximately -130 C) (10). Survival should therefore be better at very low temperatures. Confirmation was provided by Levine et al (6), who found that T. foetus survived much better on extended storage at -95 than at -28 C. There was no significant difference between these temperatures in survival up to 80 days, but thereafter the protozoa continued to die off slowly at -28, whereas their numbers remained essentially constant at -95 for 128 to 256 days, which was as long as their observations continued. The trichomonads' motility was much better after storage at -95 than after storage at -28, and fresh cultures could be initiated from the former much more readily.

References

1. Committee on Cultures, Society of Protozoologists. A catalogue of laboratory strains of free-living and parasitic protozoa (with sources from which they may be obtained and directions for their maintenance). J. Protozool. 5:1-38. 1958.

2. Diamond, L. S. The establishment of various trichomonads of animals and man in axenic cultures. J. Parasit. 43:488-490. 1957.

3. Fitzgerald, P. R., Hammond, D. M. and Shupe, J. L. The role of cultures in immediate and delayed examinations of preputial samples for Trichomonas foetus. Vet. Med. 49:409-412. 1954.

4. Fitzgerald, P. R. and Levine, N. D. Effect of storage temperature, equilibration time, and buffers on survival of Tritrichomonas foetus in the presence of glycerol at freezing temperatures. J. Protozool. 8:21-27. 1961.

5. Johnson, G. and Trussell, R. E. Experimental basis for the chemotherapy of Trichomonas vaginalis infestations. Proc. Soc. Exptl. Biol. Med. 54:245-249. 1943.

6. Levine, N. D., Andersen, F. L., Losch, M. B., Notzold, R. A. and Mehra, K. N. Survival of Tritrichomonas foetus stored at -28° and -95°C after freezing in the presence of glycerol. J. Protozool. 9:347-350. 1962.

7. Levine, N. D. and Marquardt, W. C. The effect of glycerol and related compounds on survival of Tritrichomonas foetus at freezing temperatures. J. Protozool. 2:100-107. 1955.

8. Levine, N.D., McCaul, W.E. and Mizell, M. The relation of the stage of the population growth curve to the survival of Tritrichomonas foetus upon freezing in the presence of glycerol. J. Protozool. 6: 116-120. 1959.

9. Levine, N.D., Mizell, M. and Houlahan, D.A. Factors affecting the protective action of glycerol on Trichomonas foetus at freezing temperatures. Exptl. Parasitol. 7: 236-248. 1958.

10. Meryman, H.T. Mechanics of freezing in living cells and tissues. Science 124: 515-521. 1956.

This research was supported by NIH Research Grant E-790.

GENERAL DISCUSSION

K.B. Raper, USA - I would like to comment on the preservation of slime molds. These organisms are regarded, by botanists, as slime fungi and by zoologists as mycetozoa. We can report that in the spore state, they can be lyophilized and preserved for a period of 19 years, this being the extent of our tests. This also applies to a great variety of small cyst-forming, free-living amoebae which we have encountered in our work.

I.D. Chughtai, Pakistan - Prof. Starr, what medium have you found generally suitable for transfers?

R.C. Starr, USA - If you can be content with maintaining algae contaminated with bacteria, which for many studies, except physiological ones, is perfectly adequate, the best medium is the soil-water medium developed by Dr. Pringsheim. It is a bi-phasic medium made by adding soil to distilled water and then steaming - not autoclaving - the whole works. This is really the best overall medium for isolation, for production of normal populations and for long-term storage. When you get into highly refined media, the length of time that things will last is often cut.

A. Prakash, Canada - For those of us concerned with studies of the primary production of oceans, the problem of isolation and maintenance of protozoa, bacteria and algae is quite common. This is especially so when the time of collection from the sea and the return to the base station varies from 1 to 3 months. I would like to ask Dr. Starr if there is any short-term method which can be used at sea for the preservation and recovery of viable protozoa and algae.

R.C. Starr, USA - I presume that you are referring to plankton. No, I know of nothing. Fortunately with fresh water algae we more often take mud samples at the edge of a pond rather than taking pond water, if we know that we are going to be away for more than a day. Then, the air dried soil will yield tremendous quantities of algae whenever we wish it.

E. A. George, England - You might be able to use Dr. Butcher's method. When he collects material, he puts a drop or so into a flask containing 50 to 100 ml of "Erd-Schreiber" solution. This is a way of selecting out the organisms which will survive in that medium - it is difficult, in any case, to grow the others.

PRESERVATION OF VIRUSES AND BACTERIOPHAGE

by R. E. O. Williams and E. A. Asheshov
Wright-Fleming Institute of Microbiology,
St. Mary's Hospital Medical School
and
Staphylococcal Reference Laboratory,
Central Public Health Laboratory
London, England

Preservation of viruses and phages, like preservation of other micro-organisms, may be required for several reasons. The first is the need to maintain reference stocks in a convenient fashion. For this purpose the vitally important factor is that there should be no change whatever in the behaviour of the agent; the percentage of organisms surviving the preservation process is of secondary importance, although if the percentage is low, it is very likely that some selection of variants may occur. Or preservation may be for use as an immunizing agent, as with vaccinia; the preservation of at least a sufficient proportion to ensure infection is then essential and the exact identity of the surviving agent in all its minor particulars may be less important, provided its antigenicity and, often, degree of attenuation are unchanged.

Our own interest in the preservation of staphylococcal bacteriophages has required both preservation of numbers and of exact characteristics, for we wished to be able to circulate to laboratories in all parts of the world reference samples of standard phage, to be used in comparative titrations with new stocks of phage being prepared locally.

In a general way there are two aspects of interest in the topic of preservation: the discovery of the simplest or most efficient method for preserving standard strains of the organism for reference purposes or for transmission to other laboratories; and the study of the particular factors that lead to "death" of the organism under adverse conditions of preservation and the particular metabolic activity or structure in the micro-organism that is affected. With the bacteria the stage has now been reached where the second of these topics offers the more rewarding study; in contrast, with the viruses we are still very much in the first stage. Doubtless this stems partly from the difficulty, with animal viruses at least, in making really accurate viable counts, so that the time relations of survival are difficult to assess quantitatively. At the same time there seem to be remarkably few published investigations of preservation methods; the readily availability of the deep-freeze and solid carbon dioxide seems to have very largely deflected interest from the methods of preservation so widely used for bacteria.

Animal Viruses

We are not competent to discuss in any detail the methods used for preserving animal viruses, but we can briefly review some of the literature in the hope that this may offer a framework for discussion by those who have worked actively in this field.

The need to preserve smallpox, or later vaccinia, virus for vaccination has led to many investigations of this particular virus (well reviewed by Collier (4)).

Repeated animal passage, with man as the animal, was the method used first, but very early in the history of vaccination attempts were made to preserve the virus derived from vaccination lesions on threads of cotton or between glass plates. With the introduction of "lymph" derived from lesions on animals, larger-scale preservation methods had to be introduced. The addition of glycerol to the lymph was proposed as long ago as 1850 by Cheyne, though the method was not widely adopted until 1899. The addition of glycerol to lymph is useful for reducing bacterial contamination, but does not allow prolonged survival of virus at room temperature and even at refrigerator temperature survival is limited.

Methods were developed for drying the bulk virus-containing lymph over concentrated sulphuric acid and Collier has shown that these early preparations survived at least 18 years. However, reconstitution of the virus involved grinding the desiccate, which was a difficult and dangerous proceeding.

During the 1930's Mc Clean studied the use of tissue culture virus, which proved to be very much more labile than the lymph virus, though it could be preserved at 1°C if kept quite anaerobic.

Greaves' work on the lyophilization of serum gave a great impetus to the use of this method for the preservation of microbes and Collier (5) developed a remarkably satisfactory method for freeze-drying vaccinia virus and, moreover, measured its effectiveness in proper quantitative terms. He found that survival was improved if the virus derived from sheep pulp was partially purified; suspension of the virus in 5% peptone had a substantial protective effect against the lethal action of freeze drying and subsequent prolonged storage of the dried material at high temperature. In contrast to some of the results obtained with bacteria, storage of the desiccate in a vacuum was better than storage in dry nitrogen.

Thus, within the compass of vaccinia virus we have examples of most of the preservation methods that have been used for viruses in general:-

       i) repeated animal passage
      ii) dilution with glycerol
     iii) slow drying
     iv) simple refrigeration at low temperature
      v) lyophilization.

The animal passage method need hardly be considered further: the hazards of infection to the worker and of contamination and selection of variants in the virus seem serious enough to confine its use to very few situations. The use of glycerol has been little studied, but Cabasso et al. (2) found that it has some stabilizing effect on mumps, influenza and Newcastle disease virus.

A more general review of freezing and drying for the preservation of viruses was presented by Harris (8), who drew attention again to the importance of the suspending medium, and to the fact that a medium in which viruses survive freezing is not necessarily good for their subsequent lyophilization. Solutions of gum acacia, skim milk and Lemco broth have been used with success for some viruses but not for all.

Greiff (7) has more recently presented an analysis of the process of the freezing and drying of influenza virus and showed that the best immediate preservation was attained when the virus suspension (allantoic fluid) was rapidly frozen to -76°C or -192°C. and desiccated by sublimation at 0°C or below. He considers that an important factor may be that less energy is required to abstract the water molecules when the material is vitrified than when it is in the crystalline state. Greiff also demonstrated the lethal effect of repeated freezing and thawing. None of his preservation experiments lasted more than 20-30 days.

The mode of preservation of the lyophilized material is not often considered. Harris (8) recommended storage at as low a temperature as possible, but clearly lyophilization has little advantage if refrigerator temperatures have to be employed. Lozovskaia (9) has demonstrated the deleterious effect of light on measles and influenza virus both during lyophilization and in the dried state.

With the development of a variety of new attenuated virus vaccines, the problem of preservation becomes of great importance, and it will be interesting to see how the various groups of viruses compare in their stability, and to what extent specialized techniques have to be developed for each. For practical use, the viability of the virus after reconstitution is also of great importance.

Bacterial viruses

With bacterial viruses the enumeration of the viable particles is much simpler than with animal viruses, so that quantitative studies are possible. Even so there are very few studies of different methods of preservation and practically none of the specific way in which the phage is killed by adverse circumstances. Some phages seem to be remarkably resilient: Prouty (10) obtained prolonged survival of phages from lactic streptococci dried on to filter paper at 37°C and kept in screw-capped bottles.

An early study of the preservation of phage by freeze-drying was reported by Campbell-Renton (3), who demonstrated the protective effect of broth, and obtained survival for up to 3-1/2 years. Schade and Caroline (11) also investigated freeze-drying, in this case of dysentery phages. Dialysed phage filtrates were found to be much more sensitive to the freeze-drying process than undialysed. Meat extract and egg yolk had a useful protective effect.

Preservation of staphylococcus typing phages

Our own interest derives from the need, in the Staphylococcal Reference Laboratory, to maintain a large stock of the typing phages both in fluid form ready for use in the typing of staphylococci, and in the dried form suitable for despatch as reference material to other laboratories all over the world. Latterly we have also wished to provide reference samples which could be reconstituted and titrated in parallel with each new batch of phage being prepared in national laboratories. If this is to be practicable, a method giving a really high percentage survival is essential, and, of course, the phage must preserve its range of lytic activity unchanged.

In general three methods have been used for preservation of the phages: storage at 4°C, storage at -20°C to -60°C, and lyophilization. Some years ago phages were stored mixed with their propagating strains, but since more has been discovered about the complications of lysogenization and host-induced modification, this method has been relegated to limbo.

Storage in fluid state at 4°C - It has been the policy at Colindale to maintain the stocks of typing phage filtrates at 4°C. Table I shows the titres of single batches of the 20 different typing phages tested on 2 occasions 30 months apart. Of the 20 phages 16 showed less than a ten-fold drop in titre, 3 showed about a ten-fold drop and one showed a greater drop, but still less than 100-fold.

The lytic spectra of the 20 phages were redetermined in March 1962, that is, between 3 and 4 years after the original determination. The methods have been described elsewhere (12,1). In almost all cases the patterns of lytic activity were the same; and using the coding methods described no difference was greater than plus or minus one step. There was no suggestion that even the weaker reactions were any less constant than the strong reactions.

This explicit test bears out very well our general experience over the last 10 years (12), but reports from some other workers suggest it may not be universal (e.g. 13,6). Comparable figures from other laboratories would be of great interest since it is not known what factors may have led to these apparent differences.

In an attempt to find some indication for the less satisfactory storage reported to us by some workers, the survival of 5 phages in various diluents was studied. The filtered lysates were:-
   1) M/50 phosphate buffer containing 0.001 M $MgSO_4$,
        0.0001 M $CaCl_2$ and 0.4% NaCl.

TABLE I
Survival of staphylococcal typing phages
stored in broth at 4°C

Phage No.	Titres (x 10^8)	
	June 1959	January 1962
29	3.5	1.6
52	15	1.4
52A	10	1.3
79	7.5	2.9
80	10	4.5
3A	12	6.4
3B	10	3.5
3C	12	1.4
55	3.5	0.8
71	10	14*
6	16	5.6
7	7.0	2.1
42E	8.5	1.1
47	12	3.9
53	10	0.4
54	6.4	6.1
75	1.2	0.9
77	25	12
42D	12	2.5
187	50	5.0

*This apparent increase in titre is probably
within the range of error in the titrations.

2) Nutrient broth (Oxoid).
3) Nutrient broth + 7.5% glucose.
4) Horse serum.
5) Horse serum + 7.5% glucose.
6) 5% bovine albumin + 5% sodium glutamate.

There was an immediate slight drop in titre of the phages stored
in both serum and serum + glucose, which suggested that some phage
particles were being agglutinated. Otherwise the titres remained
constant over a 6-week period and all 6 media appeared to be satisfactory as diluents for storage at 4°C.

The 5 phages were also each diluted 1:10 in nutrient broth
adjusted to various pH values between 5.8 and 7.8. The dilutions
were held at 4°C and the titre determined after 1, 7 and 14 days.
Two phages (55 and 42E) showed a significant drop in titre when
held at pH 5.8. Among the other phages (80, 75 and 42D) there was
no consistent relation between survival and pH value and the phages
retained 60% or more of their initial titre (Table II).

TABLE II

Effect of pH on survival of phages in broth at 4°C

Phage	pH	Original titre	% survival		
			after 1 day	after 1 week	after 2 weeks
80	7.8		70	85	100
	7.4		100	100	100
	7.0	1×10^9	75	90	100
	6.6		100	100	100
	6.2		75	80	100
	5.8		100	90	100
55	7.8		80	35	100
	7.4		80	100	60
	7.0	1×10^9	100	40	85
	6.6		100	65	65
	6.2		100	55	60
	5.8		100	45	25
42E	7.8		90	100	100
	7.4		100	71	73
	7.0	3.4×10^9	82	100	100
	6.7		66	85	82
	6.2		100	85	100
	5.8		32	30	23
75	7.8		80	72	60
	7.4		90	85	100
	7.0	2×10^9	90	65	100
	6.6		70	60	100
	6.2		70	60	100
	5.8		80	70	100
42D	7.8		93	60	100
	7.4		100	60	100
	7.0	4.3×10^9	100	100	100
	6.6		100	87	100
	6.2		96	90	100
	5.8		100	83	100

Storage of filtrates at different temperatures - The normal broth suspensions seem, in fact, to be reasonably stable not only at 4°C but also at -70°C and indeed at 22°C. Table III shows the results of testing 7 different phages for 16-20 weeks. There was some indication that phages repeatedly thawed and refrozen to -70°C lost titre but there was remarkably little consistent difference between the three temperatures over the period studied. One would hardly recommend 22°C as an appropriate storage tempera-ture for general use, but the phages are certainly sufficiently

TABLE III
Effect of temperature on survival of phage

Phage	Temp.	Original titre	% survival				
			1 week	2 weeks	6 weeks	16-20 weeks	Frozen and thawed*
29	-70°	5x10^10	100	100	100	52	16
	4°		100	100	100	72	
	22°		75	97	100	60	
3B	-70°	4x10^9	91	100	85	62	32
	4°		92	84	92	100	
	22°		82	86	80	40	
80	-70°	1x10^9	40			90	19
	4°		100			100	
	22°		100			20	
55	-70°	1x10^9	80			70	40
	4°		100			85	
	22°		60			10	
42E	-70°	3.4x10^9	92			73	42
	4°		67			100	
	22°		66			33	
75	-70°	2x10^9	100			45	67
	4°		80			70	
	22°		75			50	
42D	-70°	4.3x10^9	65			17	42
	4°		100			100	
	22°		80			42	

*Ampoule frozen to -70°, thawed and refrozen several times before titrating.

stable in the fluid state to be despatched by post and not to suffer as a result of any ordinary delays. Fluid phage preparations have been sent by post from Colindale to typing laboratories in Britain for several years now and no serious drops in titre have been observed.

Lyophilization - For long-term storage, and for despatch of standard phage preparations over long distances, lyophilization has obvious advantages. Again in a purely empirical way freeze-drying has been employed with success for many years at Colindale. Although occasional dryings have shown quite unexplained failures, not related to the phage, or even to the particular lysate, the results have been qualitatively highly satisfactory: the phage has been preserved and apparently quite unaltered in its lytic activity over periods of many years.

Recently some quantitative work has been done on the freeze-drying of two of the typing phages using a variety of suspending media (Table IV). For this purpose the phage was sedimented in a high speed centrifuge and the pellet resuspended. Suspension in phosphate buffer gave very poor survival on drying. With these two phages 5% bovine albumin + 5% sodium glutamate seemed to give the best protection against the lethal effects of the drying process and subsequent storage. Serum + 7.5% glucose also gave improved survival in comparison with the other media. With one of the phages the simple addition of glucose to the nutrient broth was useful. Clearly this experiment will need to be repeated on a larger scale but it gives an indication of the most profitable lines to follow. The greater part of the killing of the phage seems to occur during drying, and with almost all the preparations listed in Table IV survival of the dried material was satisfactory for several weeks at least.

TABLE IV
Comparison of suspending media for freeze-drying

Phage	Suspending medium	Titre before drying $\times 10^9$	% survival when tested Immediately after drying	1 week later	6 weeks later
29	Buffer	40	0.02	0.025	0.02
	Nutrient broth	34	7	12	6
	Nutrient broth + 7.5% glucose	31	90	70	100
	Serum	24	6	6	6
	Serum + 7.5% glucose	31	70	63	76
	5% bovine albumin + 5% Na glutamate	37	70	71	70
3B	Buffer	6	0.004	N.D.	0.0001
	Nutrient broth	6	3	"	0.7
	Nutrient broth + 7.5% glucose	6	6	"	12
	Serum	2	3	"	0.7
	Serum + 7.5% glucose	3	36	"	36
	5% albumin + 5% Na glutamate	6	70	"	55

N.D. = not done

Lyophilization of staphylococcus typing phages has also been used by Zierdt (13) who added skim milk to the lysate, and by Ghitter and Wolfson (6). In both cases the results were thought to be satisfactory.

Drying phages on paper discs - Sterile blotting paper discs, such as are used in antibiotic sensitivity testing, were impregnated with 0.02 ml of phage filtrate. The discs were then dried in vacuo over P_2O_5, either in an Edwards freeze-drying machine, which can be assumed to represent a good rapid drying (method a in Table V), or in a desiccator over P_2O_5 (method b in the Table). After drying, the discs were dropped into tubes containing 2.0 ml of broth and held at 4°C for 2 hours to allow for elution of the phage before titrating. The results are given in Table V.

TABLE V

Survival of phage dried on paper discs

Phage	Original titre	Method	% survival		
			Immediately after drying	1 week later	Usual method[*]
80	$1x10^9$	a	0.7	1.15	c 50
		b	0.04	0.03	
55	$1x10^9$	a	3.75	1.85	50-100
		b	3.75	3.8	
42E	$3.4x10^9$	a	7.0	7.3	30-50
		b	0.44	0.22	
75	$2x10^9$	a	3.2	2.8	c 80
		b	0.46	0.06	
42D	$4.3x10^9$	a	0.81	0.3	c 7
		b	0.2	0.3	

[*]Usual method = drying 0.1 ml amounts in glass ampoules.
  Method a   = desiccation in Edwards freeze drying.
  Method b   = desiccation in vacuum over P_2O_5.

As was expected, method a gave better survivals than method b but both were inferior to the usual freeze drying results, where 0.1 ml is dried in an ampoule. It may be that some of the phage is absorbed irreversibly by the paper disc; it is also possible that the type of paper used could be improved upon - the one used may contain toxic substances and very probably contains a fungicide. However the method is very quick and easy and might be useful in laboratories where the usual freeze-drying facilities are not available. It is encouraging that there does not appear to be much drop during the week of storage (at 4°C).

Other phages

In addition to staphylococcus phages, one of us (E.A.A.) has some experience of other phages. In 1958 a variety of phages were freeze-dried from broth suspensions; these included 2 staphylococcal phages, 2 streptococcal phages, 1 B. subtilis phage, 1 B. cereus phage, 1 Salm. typhi phage, 2 E. coli phages, 3 Shigella phages, 2 Pseudomonas pyocyanea phages and 2 V. cholerae phages. Losses on drying varied from 0 to 87% with an average of about 50%.

Dr. E. S. Anderson (personal communication, 1962) informs us that he has recently found lyophilization satisfactory for the preservation of the phages used for typing Salmonella typhi.

Conclusion

As more experience has been gained with the handling of bacteriophages in the laboratory, less and less difficulty seems to have been encountered in their preservation. Most of the phages tested seem to be remarkably tough and able to withstand prolonged storage in broth at 4°C or lyophilization. Precise studies of the mode of death during lyophilization do not seem to have been reported.

Animal viruses appear to be much more variable in their behaviour under various methods of preservation but there is clearly a great need for more comparative studies in which quantitative assessment is made of survival through the different stages of preservation.

References

1. Blair, J. E. and Williams, R. E. O. Phage typing of staphylo-cocci. Bull. WHO 24: 771-784. 1961.
2. Cabasso, V. J., Markham, F. S. and Cox, H. R. Stabilizing action of glycerine on haemagglutination of egg-adapted mumps, Newcastle disease and influenza viruses. Proc. Soc. Exp. Biol. Med. 78: 791-796. 1951.
3. Campbell-Renton, Margaret L. Experiment on drying and on freezing bacteriophage. J. Path. Bact. 53: 371-384. 1941.
4. Collier, L. H. The preservation of vaccinia virus. Bact. Rev. 18: 74-86. 1954.
5. Collier, L. H. The development of a stable smallpox vaccine. J. Hyg. (Lond.) 53: 76-101. 1955.
6. Ghitter, L. R. and Wolfson, S. W. A simplified approach to the phage typing of Staphylococcus aureus. I. The use of lyophilized phage. Amer. J. Clin. Path. 34: 77-91. 1960.
7. Greiff, D. The effects of freezing, low temperature storage and drying by vacuum sublimation on the activities of viruses and cellular particulates. In Recent research in freezing and drying. Edited by A. S. Parkes and Audrey V. Smith. Blackwell, Oxford. 1960.
8. Harris, R. J. C. The preservation of viruses. In Biological applications of freezing and drying. Edited by R. J. C. Harris. Academic Press, New York. 1954.
9. Lozovskaia, L. S. Destructive action of light on measles and influenza viruses during desiccation and subsequent storage in vacuo. Probl. Virol. 4: 56-59. 1959.

10. Prouty, C. C. Storage of the bacteriophage of the lactic acid
    streptococci in the desiccated state with observations on
    longevity. Appl. Microbiol. 1: 250-251. 1953.
11. Schade, A. L. and Lerna, Caroline. The preparation of a poly-
    valent dysentery bacteriophage in a dry and stable form. I.
    Preliminary investigations and general procedures. J. Bact.
    46: 463-473. 1943. II. Factors affecting stabilization of
    dysentery bacteriophage during lyophilization. Ibid. 48: 179-
    190. 1944.
12. Williams, R. E. O. and Rippon, J. E. Bacteriophage typing of
    Staphylococcus aureus. J. Hyg. (Lond.) 50: 320-363. 1952.
13. Zierdt, D. H. Preservation of staphylococcal bacteriophages
    by means of lyophilization. Amer. J. Clin. Path. 31: 326-331.
    1959.

DISCUSSION I

by R. L. Thompson
National Cancer Institute
Bethesda, U. S. A.

Dr. Williams has referred to the need for preservation of
reference strains of viruses. I would like to describe briefly a
collection of reference strains of viruses and rickettsiae which is
maintained at the American Type Culture Collection in Washington,
D. C. This is the Viral and Rickettsial Registry.

The Registry was established in 1949. The objectives were
first to insure the continued existence of prototype or reference
strains of viruses and rickettsiae of human or animal origin and
second to provide an efficient means for the distribution of these
agents to scientists and teachers having need of them. Strains in
the following categories now are accepted for inclusion in the
Registry:

1) Classical strains.
2) Reagent strains, including items deposited by the
   National Institutes of Health.
3) Attenuated strains.
4) Early passage levels.
5) Special items of value for specific purposes.

Each strain deposited in the Registry is fully documented by
the investigator who supplies it. This documentation has been published in Registry catalogs (1st Ed., 1950; 2nd Ed., 1959; Suppl.,
1961). At present the collection contains approximately 270 strains.

The Registry is controlled by the Viral and Rickettsial
Registry Committee. This committee is composed of contributors

to the collection and of a few other interested individuals. The committee designates certain of its members to serve on an Executive Council which is responsible for the implementation of policies concerning the operations of the organization. The American Type Culture Collection has acted as a depository and distribution agency for the Registry since it was established. In 1962, the American Type Culture Collection distributed over 1,000 items from the collection.

It has been a basic policy of the Registry Committee that the reference stock of each agent in the collection should be prepared by a competent, experienced investigator in his own laboratory. When the stock of a given agent reaches a low level, the donor is asked to prepare a new lot, using seed from the earlier lot. If he is unable to do so, the Executive Council will select another investigator to assume this responsibility. Since the demand for certain strains is limited, reassays of these stocks are made when indicated. As a rule, the reassays are made by the original donor.

The reassay of stocks of strains in the Viral and Rickettsial Registry have provided information with regard to the survival of these agents under standard storage conditions. Data for several strains are presented in Table I.

Some of the stocks were prepared in eggs and others in mice. All were lyophilized and stored at -20°C. It will be noted that the Henzerling strain of Q fever rickettsia still had a good titer after storage for 10 years. The Iowa #15 strain of swine influenza virus likewise was viable after storage for 10 years. The titer of the 17D strain of yellow fever virus decreased 1.7 log units after storage for 7-1/2 years.

With the exception of the stock of Japanese encephalitis virus, all viral stocks in mouse brain tissue showed some reduction in titer after storage for over five years.

The demand for a number of items in the Registry has been such that long-term data for viability of these strains has not been obtained. Most recent additions to the collection have been received in liquid form and these have been stored at -60°C.

Facilities are now available at the American Type Culture Collection for the storage of materials in liquid nitrogen. The precise manner in which these facilities are to be utilized for the preservation of rickettsiae and viruses remains to be determined.

TABLE I

Viability of Lyophilized Stocks of Rickettsiae and Viruses
stored at -20°C

ATCC No.	Agent	Strain	Titer Original	Reassay	Time (mo.)
		Stocks Prepared in Eggs			
142	Epidemic typhus	Breinl	-	*	66
144	Murine typhus	Wilmington	-	*	66
145	Q fever	Henzerling	-	4.3	120
125	Psittacosis	6BC	6.9	6.3	21
124	Ornithosis	P-4	3.5	3.2	48
122	Meningopneumonitis	Francis	7.5	4.5	32
123	Mouse pneumonitis	Nigg II	4.5	4.5	56
120	Feline pneumonitis	No. 1	3.8	4.8	26
23	Laryngotracheitis	Lederle	4-5	4.0	74
99	Influenza (swine)	Iowa #15	2.0	*	120
114	Fowlpox	Beaudette	8.0	*	27
118	Vaccinia	N.Y.B.H.	8.0	8.0	37
83	Yellow fever	17D	7.0	5.3	89
		Stocks Prepared in Mice			
85	Anopheles A	Original	4.6	4.2	67
86	Anopheles B	Original	6.0	4.4	67
88	Bwamba	Original	6.3	4.8	79
73	Ilheus	Original	8.8	7.3	66
74	Japanese Encephalitis	Nakayama	7-8	7.7	104
78	Ntaya	Original	4.9	4.5	67
82	West Nile	B 956	-	3.8	78
91	Wyeomyia	Original	6.6	4.3	66
84	Zika	MR 766	5.8	4.8	66
134	LCM	Armstrong	-	3.6	72

* Viable

DISCUSSION II

by R. Wahl
Institut Pasteur
Paris, France

Some observations pertinent to the conservation of bacteriophages will be outlined here.

Conservation in a liquid medium

For the conservation of viruses and phages in a liquid medium protective substances (albumin or peptones) are necessary. Phages lysates, when preserved for several months at +4°C in peptone broth at neutrality keep their level of infectivity. Certain phages like S_{13} are sensitive to light (9). Oxidation is a major contributory factor to inactivation especially in non-protective media. In the absence of air, S_{13} and C_{16} phages can be conserved for a long time in a saline medium containing sodium hydrosulfite (10). Adams (1) attributes the inactivation of viruses in saline medium to reactions at the interphase.

Conservation of viruses in the frozen state, at a very low temperature gives good results (Harris (4)) provided that successive freezing and thawing are avoided.

With lyophilisation, three stages must be considered:

a) Rapid freezing should occur at a critical temperature, which may vary between -30°C to -40°C, and ever lower according to the medium. There is no well established interpretation of this fact: end of the eutectic point, modification of the crystalline form etc...

b) The desiccation of the frozen sample takes place in two phases:

During the sublimation of the ice the temperature of the sample should be inferior to its own critical temperature, and yet high enough for the sublimation to be as rapid as possible. To achieve this, the calories lost during sublimation must be rigorously replaced at each instant, and the pressure regulated. Bauer and Pickel (2) had already noted that yellow fever virus is inactivated when the medium liquefies during lyophilisation.

To accomplish secondary desiccation the temperature must be raised to +35°C or +40°C.

c) Storage necessitates particular conditions.

Two types of lyophilisation are possible:

a) Samples of at least 2.5 to 3 ml are lyophilised in pellet-form in a bell-shaped apparatus, after determination of the critical temperature from the resistivity curve during thawing (Rey (6)). With this type of lyophilization, the temperature and pressure can be automatically controlled during the two phases (Rey (6)).

b) Lyophilisation in thin films of 0.5 to 0.1 ml in an apparatus "en hérisson" (with numerous side-arms) was carried out in an air conditioned room with rapid air turnover. The thin film which permits a rapid exchange of heat is obtained by horizontal mechanical rotation of the tubes in a bath at -70°C. In spite of the data obtained from the resistivity curve, the temperature of the room during the first phase had to be obtained empirically for each type of operation. For example, the temperature is around +10°C for yellow fever virus during the first phase; it is raised to +35°C for the second (5).

While lyophilisation in thin films is the only technique which reduces the risk of contamination, lyophilisation in pellet-form is useful in determining the best conditions of lyophilisation.

Effect of substances present in the lyophilisation medium

Certain apparent contradictions in the literature would seem to be due to different experimental conditions. Vieu (7,8), Wahl and Fouace (11) using the conditions outlined above for a certain number of phages have obtained the following results:

a) The effect of substances suitable or not for preservation is more marked on centrifuged phages than on phage lysates.

b) The proportion of phages destroyed does not depend on the initial titre, but only on the medium.

c) Non-suitable substances are: those which decrease the freezing point (e.g. glycerol, gelatin, agar), those responsible for too much residual humidity after drying, e.g. glucose, serum-albumin; the latter are noxious for viruses (Harris (4)) but favourable for bacteria. Sodium chloride and mannitol are noxious. Peptone broth (without NaCl), beer wort and sucrose are the best preservatives. No attempt was made with gum acacia as recommended by Campbell-Renton (3). Skim milk, sodium glutamate, dextran, subtosan are without effect.

d) Certain very fragile phages, such as the Twort phages of Staph. aureus can be preserved in suitable media when lyophilisation is carried out with extreme precaution; while others (e.g. T_2 and $\emptyset X$) are less sensitive. The titre of phages 42D and 29 of Staph., as well as Vi II, a phage of Proteus showed relatively little decrease even after inadequate lyophilisation.

The conservation of viruses during storage has not been extensively studied. It necessitates a more thorough desiccation than that used for bacteria, and for certain viruses, e.g. yellow fever a temperature of storage of -20°C. Lyophilised phages can be stored at ordinary temperatures.

References

1. Adams, M.H. The surface inactivation of bacterial viruses and proteins. J. Gen. Phys. 31:417. 1948.

2. Bauer, J.H. and Pickel, E.G. Apparatus for freezing and drying in large quantities under uniform conditions. J. Exp. 71: 83. 1940.

3. Campbell-Renton, M.L. Experiments on drying and freezing bacteriophage. J. Path. Bact. 53:371. 1941.

4. Harris, R.J.C. The preservation of viruses. In Biological applications of freezing and drying. Acad. Press, New York. 1954.

5. Panthier, R. Présentation d'appareils utilisés pour la préparation du vaccin antiamaril. Bull. Soc. Pathol. Exotique 49: 616. 1956.

6. Rey, L.R. Thermal analysis of eutectics. In Freezing and drying materials. Ann. N.Y. Acad. Sci. 85: 510. 1960.

7. Vieu, J. F. Sur la lyophilisation des bactériophages. C. R.
     Acad. Sci. 252: 1230. 1961.
8. Vieu, J. F. and Diverneau, G. Lyophilisation du bactériophage
     Vi II. C. R. Acad. Sci. 254: 3149. 1962.
9. Wahl, R. and Latarjet, R. Inactivation de bactériophages par
     les radiations de grande longueur d'onde. Ann. Inst. Pasteur
     73: 937. 1947.
10. Wahl, R. and Blum-Emerique, L. Conservation du bactério-
     phage et potentiel d'oxydo-réduction. Ann. Inst. Pasteur
     72: 959. 1946.
11. Wahl, R. and Fouace, J. unpublished.

GENERAL DISCUSSION

B. Babudieri, Italy - Dr. Thompson, three questions: 1) How are
the trachoma virus strains preserved? 2) In preserving Rickettsia
or large viruses, do you lyophilize purified suspension of the virus
or do you start with yolk-sac material or arthropods? 3) Have you
observed that it is not necessary to use lyophilization for long
term preservation of rickettsial strains but only to dry infected
arthropods?

R. L. Thompson, USA - We have not added trachoma strains to our
collection as yet because of some uncertainty as to which would be
the logical ones to keep. At the meeting in Montreal, the Trachoma
Study Group elected to add some representative strains. Presumably all of these will be for frozen storage, not lyophilization. The
rickettsial strains are all yolk-sac material and have been quite
satisfactory.

B. Blaskovic, Czechoslovakia - We have experienced no difficulty
in our institute in maintaining the myxoviruses which reproduce
very well in chick embryos. They can be stored for 5 to 6 years
when frozen and dried. The other group of viruses which we maintain are the arbor viruses. Our strains are maintained in 5 or 10%
mouse brain suspension and can be stored at -40°C for a long
time. Another possibility, which has not been investigated, would
be to maintain viruses belonging to the tick-borne group, in ticks,
i.e. in a very natural condition when the ticks are maintained at
+4°C or 0°C.

R. L. Thompson, USA - There is great variation in the resistance
of viruses and it is surprising how long some strains do persist
under rather ordinary circumstances. However, when we begin to
clean up strains - as we begin to remove protein - we get into
more and more trouble with the various agents. This, then, is one
of our problems.

PRESERVATION AND CHARACTERIZATION OF ANIMAL CELL STRAINS

by C. S. Stulberg
The Child Research Center of Michigan
Detroit, U.S.A.

Introduction

In 1960, under the sponsorship of a U.S. federal agency, the National Institutes of Health, the Cell Culture Collection Committee was formed to initiate and coordinate a program for characterizing and preserving animal cell strains, and to establish a repository and distribution center for reference cultures. The primary aims of this program include a systematic characterization and stand-ardization of prototype cultures and their variants, development of criteria for certifying strains, and distribution of such strains to all qualified investigators. The program was made operationally feasible by advances in cryobiology, recently reviewed in a mono-graph by Smith (15), and the specific application of freezing and ultra-low temperature techniques to the preservation of cultured animal cell strains by a number of investigators (3, 5, 8, 9, 12, 13, 16-18).

The choice of operating procedures was dependent on the cell-strain properties that must be preserved. Two problems were immediately apparent. The first involved the development of freeze-preservation methods suitable for preparation and storage of reference stocks, wherein reasonable numbers of viable cells could be recovered, unaltered in their properties and repropagated within a relatively short period of time. The estimate of efficiency of such procedures would depend on both quantitative and qualita-tive criteria used for judging the recovery of viable cells from the frozen state, thus allowing the determination of optimal freezing and thawing rates, freeze media, storage temperatures, and attendant handling operations.

The second, and more difficult problem was that of cell characterization itself. Means were available for determining: (a) morphology under defined culture conditions; (b) growth char-acteristics including nutrient requirements, mean generation time, lag and log phase growth characters, and plating efficiencies; (c) chromosomal characterization with designation of known markers, and with analyses of chromosomal frequencies; (d) biochemical genetic markers, such as resistance to certain purine analogs; (e) immunologic specificity with relation to species of origin; (f) tumorigenicity by homograft or heterograft reactions; and (g) sus-ceptibility to selected viruses. Inadequate as such properties might be for classification or even identification of cell strains, they at least provide an approach toward a systematic characterization. It is obvious, therefore, that any freeze-preservation procedure used in a cell culture collection must be designed to return optimal numbers of viable cells without risk of selection for only certain recognizable characteristics.

Practical Considerations of Freeze-Preservation of Cell Strains

A number of studies (3,5,8,9,12,13,16,17) have demonstrated the practicability of preserving animal cell strains by utilizing slow cooling in glycerol-containing media, storage at temperatures below -70°C, and recovery by rapid thawing. Although this general procedure usually returned viable cells, deviations in freezing rates, glycerol concentrations, and storage temperatures, as well as differences in criteria used for judging the efficiency of recovery from the frozen state, prevented a strict comparison of results. Nevertheless, while satisfactory explanations of the physical-chemical mechanisms involved are not yet available (11), the forementioned reports did suggest empirical approaches toward the establishment of a useful preservation system for cultured animal cells (18). These might be evaluated by considering two aspects of preservation, i.e., the effects of freezing and thawing procedures, and the effects of storage temperature.

Most data indicated that slow cooling at controlled rates (usually 1-2°C per minute), in the presence of a freely penetrating protective agent (glycerol) produced good results. Although cooling rates up to 45°C per minute (but not higher) were stated to be effective for two cell strains (9), it has been pointed out that the faster cooling rates often are not optimal, and in no event have they produced better results. While optimal concentrations of glycerol were not established, it was evident that concentrations as low as 5% in the freeze medium afforded maximal protection for a variety of cell strains (16-18). The latter concentration of glycerol did not require its separate removal after thawing, provided that the cell density was such that the sample could be sufficiently diluted (approx. 1:10) before culturing. Recently it has been shown (9) that 5% dimethylsulfoxide appears to give results comparable to 5% glycerol, but in addition has the possible advantage of permeating the cells more rapidly.

Evidence that the rate of thawing has a marked effect on cell recovery has been clearly demonstrated both with red cells (11) and with mammalian cell strains (20). Such results indicate that thawing of suspensions of cultured cells should be accomplished as rapidly as possible.

It can be assumed a priori that the lowest storage temperature practicably obtained should be used for long term storage. Storage temperatures obtained with dry ice were considered to be more or less adequate in earlier studies, where cells could readily be re-established in culture after being maintained for several years at such temperatures (12,16,17). More recent quantitative data (20) however have indicated that an appreciable decay in cell viability occurs over a long period at temperatures of -75 °C. Liquid nitrogen apparatus, maintaining constant temperatures as low as -195 °C is now commercially available and provides optimum conditions for storage of biological materials.

Freeze-Preservation Methodology for an Animal Cell Culture Collection

The foregoing considerations formed the basis for procedures that were developed and put into practice in the program for characterization and permanent preservation of cultured cell strains. Following is a description of our current preservation techniques and examples of maintenance of mammalian cell strain characteristics (18). It was necessary to be able to preserve from each strain enough representative cell samples to meet the needs of characterization studies as well as to obtain a suitable quantity for a permanent repository. This entailed (a) factors pertaining to the preparation of cells, (b) uniform freezing, ultra-low temperature maintenance, and thawing operations, (c) quantitative and qualitative determination of viability, growth, and other cell characteristics as affected by (b).

For these purposes, cell strains to be initially included in the collection were cultivated as monolayers in basal media supplemented with serum (according to the requirements of each cell strain). Large bottle cultures were employed to yield suitable numbers of cells for a reference seed stock. After appropriate incubation, the cells were freed from the monolayers with trypsin according to standard methods, and resuspended in the "freeze" medium which consisted of Eagle's basal medium 80%, serum 15%, and glycerol 5%. A typical suspension of cells prepared in this fashion contained approximately 10^9 cells. About 5×10^6 cells were dispensed into 1 ml thick-walled ampules, glass-sealed, and the resulting 200-300 ampule batches were frozen in an automatic controlled-rate slow freeze unit (employing liquid nitrogen vapor) and stored in a liquid nitrogen refrigerator. Details of this procedure together with the slow freezing and rapid thawing rates employed will be found in reference (18).

Effects of Freeze-Preservation on Cell Viability

An assessment of the efficiency of the described preservation procedure is provided by two indicators of viability, dye exclusion and capabilities of individual cells to attach to surfaces and to multiply (plating efficiency). The term "viable cells" is arbitrarily equated with cells that were unstained with trypan blue, and plating efficiencies are expressed as the ratio of the number of cell colonies times the dilution: number of unstained cells per ml.

Table I compares the viabilities and plating efficiencies of a variety of cell strains just prior to freezing and following thawing after a storage interval of 3 months. (Tests after one year's storage revealed no change). Viability counts on all strains before freezing were essentially similar, while the plating efficiencies were characteristic of the strain. Upon thawing, viability counts revealed that practically no loss had occurred with some (similar) strains, and a relatively small loss with others, but in all cases, the degree of survival appeared to be a function of the particular strain under these conditions. Plating efficiencies of the recovered viable cells were essentially the same as those frozen (variations are within the range of reproducibility) which indicated that regardless of the loss in viability, their plating characteristics were retained. This index of viability, indicates that trypan blue staining can serve as a quantitative control of preservation processes, whereas plating efficiency can serve as a qualitative control of selection pressures with regard to this one characteristic.

TABLE I

Comparative viability data of several cell strains
stored at low temperatures*

Cell strains	Cell type	Prefreeze viability determinations		Average recovery of cells from storage	
		Viable cells % unstained	Plating efficiency %	Viable cells %	Plating efficiency %
Hela 229	Human, Ep-L	98.3	36.4	95.7	37.8
Detroit-6	Heteroploid	96.9	53.8	96.7	55.0
Detroit-504	Human, Fb-L Diploid	95.5	16.7	89.5	9.3
S-180 11	Mouse-Tum-origenic	94.2	65.4	85.0	59.6
LLC-MK2	Monkey-Ep-L	98.7	17.3	84.5	13.2

*After Stulberg et al. (18)

There has been considerable interest in the preservation of cells of primary cultures of various types of tissues. Although such cells are not readily amenable to plating studies, trypan blue staining indicates a relatively high recovery of presumably viable cells. Table II illustrates recovery data with primary cultures of cells derived from kidney tissues of a variety of species.

These viability counts were correlated with the ability of the recovered cells to attach to glass and form complete monolayers with growth curves equivalent to unfrozen cells.

TABLE II

Recovery of cells of primary cultures after freeze-preservation

Source of primary cells	Prefreeze % viable	Post-freeze % viable
Human kidney	95-98	76-81
Monkey "	78-96	60-90
Mouse "	86-96	52-62
Rabbit "	83-91	62*

*Result of one determination

Effect of Freeze-Preservation on Other Characteristics of Cultured Cells

Unpublished data has accumulated in a number of laboratories that suggest that most, if not all, currently identifiable characteristics of cultured cells will survive freeze-preservation provided reasonable attention is given to the control of factors contributing to cell injury or death.

For example, general morphologic and growth characteristics in cultures of thawed cells indicate that such properties are easily preserved. In addition, there appears to be no adverse effects of freeze-preservation on nutritional requirements of cultured cells. Usually, the medium used as a freezing vehicle has been of the same composition as that used for cell propagation plus the optimal concentration of glycerol, and in some instances a small increase in the amount of serum. However, the development of cell strains capable of growth in chemically defined media has raised the question whether or not such strains can be efficiently preserved in the absence of serum or other proteins in the media. There has been some indication that absence of serum may markedly reduce the recovery of viable (unstained) cells, of a strain capable of growth on synthetic media (18). However, more recent studies (4,14) of three different strains adapted to growth on chemically defined media have shown that such cells, preserved in as little as 6% glycerol in the absence of protein, can be readily re-propagated with no change in nutritional requirements or growth rates.

Curiously, while the chromosomal complement of cultured cells is obviously one of the more significant characteristics to be preserved, very little published information is available on this application. General experience in the Cell Culture Collection program, however, and that of other investigators has been that no changes in karyotypes can be detected due to freeze-preservation. This is a finding of particular importance with regard to maintenance of diploid strains (7).

Genetically marked cell strains characterized by their

resistance to purine analogs (22), species-related surface antigens (1,2,19), virus susceptibilities (17), and tumorigenicity by homograft (10) and heterograft (6) reactions, are additional examples of specific characteristics of cultured cells that are retained through freeze-preservation.

*   *   *   *   *   *   *

Hence, applications of freezing and ultra-low temperature storage techniques to the preservation of mammalian cells have made feasible the establishment of a culture collection of animal cells. Some current aspects of quantitative and qualitative assessment of preserved cells have been briefly outlined.

Although much of the foregoing methodology also applies to preservation of cells of transplantable tumors, a discussion of the latter is beyond the scope of this paper. Reference is made to Hauschka et al. (8) and Kline et al. (10) concerning cold storage banks of tumors and associated problems.

References

1. Brand, K. G. and Syverton, J. T. Immunology of cultivated mammalian cells. I. Species specificity determined by hemagglutination. J. Nat. Cancer Inst. 24: 1007-1019. 1960.

2. Coombs, R. R. A. Identification and characterization of cells by immunologic analysis, with special reference to mixed agglutination. In Analytic Cell Culture. National Cancer Institute Monograph #7. Edited by R. E. Stevenson. U.S. Government Printing Office, Washington, D. C. 1962.

3. Craven, C. The survival of stocks of HeLa cells maintained at -70°C. Exp. Cell Res. 19: 164-174. 1960.

4. Evans, V. J., Bryant, J., Montes-DeOca, H., Schilling, E. and Shannon, J. E. Liquid nitrogen freezing of cells in chemically defined medium. (Abstract) 13th Ann. Meeting Tissue Cult. Ass'n., Washington, D. C., May 29-31, 1962.

5. Ferguson, J. Long term storage of tissue culture cells. Australian J. Exp. Biol. 38: 389-394. 1960.

6. Foley, G. E., Handler, A. H., Adams, R. A. and Craig, J. M. Assessment of potential malignancy of cultured cells: Further observations on the differentiation of "normal" and "neoplastic" cells maintained in vitro by heterotransplantation in Syrian hamsters. In Analytic Cell Culture. National Cancer Institute Monograph #7. Edited by R. E. Stevenson. U. S. Government Printing Office, Washington, D. C. 1962.

7. Hayflick, L. and Moorhead, P. S. The serial cultivation of human diploid cell strains. Exp. Cell Res. 25: 585-621. 1961.

8. Hauschka, T. S., Mitchell, J. T. and Niederpruem, D. J. A reliable frozen tissue bank: viability and stability of 82 neoplastic and normal cell types after prolonged storage at -78°C. Cancer Res. 19: 643-653. 1959.

9. Kite, J. H. and Doebbler, G. F. Effects of cooling rates and additives on survival of frozen tissue culture cells. (Abstract) Fed. Proc. 20: 149. 1961.

10. Kline, I., Acker, R. F., Anderson, G. and Schepartz, S. Preservation and characterization of cultured cells and animal tumors. Cancer Chemotherapy Reports, in press, 1962.

11. Meryman, H. T. Freezing of living cells: Biophysical considerations. In Analytic Cell Culture. National Cancer Institute Monograph #7. Edited by R. E. Stevenson. U.S. Government Printing Office, Washington, D.C. 1962.

12. Scherer, W. F. and Hoogasian, A. C. Preservation at subzero temperatures of mouse fibroblasts (strain L) and human epithelial cells (strain HeLa). Proc. Soc. Biol. & Med. 87: 480-487. 1954.

13. Scherer, W. F. Effects of freezing speed and glycerol diluent on 4-5 year survival of HeLa and L cells. Exp. Cell Res. 19: 175-176. 1960.

14. Shannon, J. E., DenBeste, H. and Evans, V. J. Experiments on the freezing of cell lines in liquid nitrogen vapor. (Abstract) 13th Ann. Meeting Tissue Culture Ass'n. Washington, D.C. May 29-31, 1962.

15. Smith, A. U. The Biological Effects of Freezing and Supercooling. Williams & Wilkins Company, Baltimore. 1961.

16. Stulberg, C.S., Soule, H.D. and Berman, L. Preservation of human epithelial-like and fibroblast-like cell strains at low temperatures. Proc. Soc. Exp. Biol. & Med. 98: 428-431. 1958.

17. Stulberg, C. S., Rightsel, W. A., Page, R. H. and Berman, L. Virologic use of monkey kidney cells preserved by freezing. Proc. Soc. Exp. Biol. & Med. 101: 415-418. 1959.

18. Stulberg, C. S., Peterson, W. D., Jr. and Berman, L. Quantitative and qualitative preservation of cell strain characteristics. In Analytic Cell Culture. National Cancer Institute Monograph #7. Edited by R. E. Stevenson. U.S. Government Printing Office, Washington, D.C. 1962.

19. Stulberg, C. S., Simpson, W. F. and Berman, L. Species-related antigens of mammalian cell strains as determined by immunofluorescence. Proc. Soc. Exp. Biol. & Med. 108: 434-439. 1961.

20. Stulberg, C. S. Unpublished data.

21. Swim, H. E., Haff, R. F. and Parker, R. F. Some practical aspects of storing mammalian cells in the dry-ice chest. Cancer Res. 18: 711-717. 1958.

22. Szybalski, W., Szybalska, E. H. and Ragni, G. Genetic studies with human cell lines. In Analytic Cell Culture. National

Cancer Institute Monograph #7. Edited by R. E. Stevenson.
U. S. Government Printing Office, Washington, D. C. 1962.

DISCUSSION I·

by J. F. Morgan
University of Saskatchewan
Saskatoon, Canada

Dr. Stulberg, in his excellent presentation, has outlined the
aims of the Cell Culture Collection Committee and has described the
present status of the knowledge which this Committee has accumulated. Of the two major problems which Dr. Stulberg has outlined,
it does appear that the first one -- the development of suitable
methods for freeze-preservation and recovery of viable cells -- has
been largely overcome. It is now possible to freeze a large variety
of mammalian cell types, preserve them by ultra-low temperature
techniques, and recover reasonable numbers of viable cells. As a
consequence, the establishment of a culture collection of animal
cells has now become feasible. This is a major advance for workers
in the cell cultivation field, since mammalian cells present the
same problems of variability and alteration in characteristics exhibited by bacteria and other species of micro-organisms.

The second problem specified by Dr. Stulberg -- that of cell
characterization -- presents many more difficulties which are not
susceptible to easy solution. These difficulties reflect in large part
our very fragmentary knowledge of mammalian cells, of their
nutrition, their metabolism, their genetic constitution, their enzyme
content. Detailed studies on all these characteristics are urgently
needed before we can be certain that mammalian cells preserved
through ultra-low temperature storage have, in fact, retained their
basic characteristics.

A major difficulty in preservation studies with cell cultures
lies in the fact that the cell strains presently available appear to
have somewhat different nutritional requirements which necessitate
the supplementation of any basal medium with varying types and
quantities of animal serum. This inherent source of variability in
the method would be overcome if a chemically-defined medium,
adequate for all cell strains, could be devised.

Dr. Stulberg has stated that the degree of survival appears to
be a function of each particular strain. The reasons for this variability require further investigation. It is possible that detailed
nutritional studies may reveal the presence of small molecular
protective factors which could be added to the culture medium to
increase subsequent cell viability.

In the techniques summarized by Dr. Stulberg, mammalian cells for preservation are released from their monolayers by treatment with trypsin. A considerable body of evidence has accumulated that such treatment may have deleterious effects upon the membranes of cells in culture. It would be important to know whether the trypsin procedure affects the subsequent viability after preservation. Possibly a less rigorous method of releasing the cells from their monolayers might be devised.

In a previous paper in this conference, Dr. Bradley has discussed the loss of adaptive enzymes during storage of yeast cells and has raised the possibility that subtle changes in the genetic mechanism may occur. That similar changes may occur in mammalian cells is illustrated by studies reported by my colleagues, Mr. L. F. Guerin, Miss H. J. Morton, and myself, employing mouse ascites tumor cells as our test system (1). We suspended the washed tumor cells in modified Tyrode's solution containing 10% glycerol, froze them in a dry ice-ethanol bath, stored them over dry ice for varying periods of time, and revived them by rapid thawing. Before this treatment, two ascites tumor lines were strain specific, producing tumors only in the mouse strains of origin. After the preservation treatment, these two tumor lines were found to have lost their strain specificity and would now produce tumors in any mouse strain tested. This loss of mouse strain specificity appears to have been irreversible since it has now been maintained for more than six years in our laboratory. While these results may not be directly applicable to the situation with mammalian cell cultures, they do emphasize, as Dr. Stulberg has done, the urgent need for further detailed studies on the identifiable characteristics of mammalian cells and their response to low-temperature preservation.

References

1. Morgan, J. F., Guerin, L. F. and Morton, H. J. The effect of low temperature and storage on the viability and mouse strain specificity of ascitic tumor cells. Cancer Res. 16: 907-911. 1956.

DISCUSSION II

by Robert E. Stevenson
National Cancer Institute
Bethesda, U.S.A.

Dr. Stulberg has succinctly reviewed the current thoughts and methodology directed toward the preservation and characterization of animal cell strains. He has also pointed out that classification and identification of cell lines is a problem of a different order of magnitude. Although no solutions to this problem are expected in the near future, it might be advisable to consider the practical implications of this handicap for the establishment of cell culture collections.

With other forms of organisms, with the possible exception of viruses, it has been found possible to establish sufficient species criteria for recognition. It seems probable that with animal cells the species from which they were derived can be determined by various immunologic techniques. One therefore has major breakdowns into human, mouse, bovine, etc., cells. When it comes to definition of cell types within the species, however, we then find that our criteria are inadequate for separation into clearly recognizable cell types.

To further compound the problem, "transformations" occur in the course of serial cultivation of cell lines. As Sanford (1) has shown with derivatives from a single mouse cell, lines with radically different properties may be selected and maintained over a period of time. In some cases, as Scherer (2) has pointed out, the salient desirable feature of a particular cell line can be lost as was shown with some derivatives of the HeLa strain which fail to support the growth of polio virus. These observations re-emphasize the need for thorough characterization of cell lines to the best of our ability, employing all known useful techniques for establishing biological properties as Dr. Stulberg has enumerated. By such careful control, it is possible for the individual investigator to select and employ a cell line possessing characteristics which are necessary for his experiments.

The curator of cell collections, on the other hand, is faced with a task of identifying cell lines which are useful to the scientific community and which are indeed the ones described in the original publications on the line's isolation. If inadequate records have been kept of the genealogy of a particular cell line or if the cell line has been received second or third hand from the original investigator, it is scientifically impossible at the moment to state without doubt that such a cell line is indeed representative of its "prototype".

Another source of confusion is the lack of standardized nomenclature for describing or identifying cell lines. This problem has been examined by at least two groups. The International Meeting on Tissue Culture which took place in Glasgow, Scotland, in 1957 recommended a system of letters and numerals which would indicate the laboratory of origin of the cell lines and identification of the line by its unique number.

Currently, differences of opinion on the naming of cell lines versus cell strains based upon their biological implications have

impeded the adoption of any standardized system. Accordingly, the
Cell Culture Collection Committee has decided to identify cells in
its program by catalogue number followed by the complete description of the known biological properties and genealogy of the
strain. This method is temporizing at best and suffices only since
confusion exists.

Before complete chaos on an international scale exists in
cell culture nomenclature, it is to be recommended that curators
of cell collections form a bond of common need and mutually agree
on methods for retaining some order and identification for their
collections. Such an arrangement would have the advantage of
making some degree of biological standardization available on an
international basis provided by exchange of seed cultures which
are thoroughly characterized. Newer acquisitions by any one collection could be readily identified for the others so that duplicate
terminology could be avoided.

Hopefully, we shall look forward to a day when it is possible
to determine whether all cell lines isolated from one tissue are
identical in their properties or whether significant differences
exist. Until such time as we are able to perform such definitions,
some means will be necessary to avoid confusion and the implica-
tion that we know more than we actually do. A well organized collection of cell lines would make possible extensive comparisons of
cells based on many parameters.

The time is long past since a curator of cell lines, bacteria,
viruses or any other material should be equated with a guardian of
a bank vault. Just as a curator of an art collection has responsibility for enlarging and balancing various items in the collection
and educating the public to meaningful comparisons, the biologist
in a comparable role should also meet this challenge.

References

1. Scherer, William F. Comparative susceptibility of cells of the
      same type to infection by poliomyelitis virus. Ann. New York
      Acad. of Sciences Vol. 61, Art. 4: 806-821. 1955.
2. Sanford, K. K., Likely, G. D. and Earle, W. R. The develop-
      ment of variations in transplantability and morphology within
      a clone of mouse fibroblasts transformed to sarcoma-
      producing cells in vitro. J. Natl. Cancer Inst. 15: 215-237.
      1954.

GENERAL DISCUSSION

J.A.R. Miles, New Zealand - I would like to draw attention to a
point which was raised by Porterfield earlier this year. He reported that dimethyl sulfoxide was a great improvement on

glycerol as an additive for preserving mammalian cells. We have never in the past been able to get such good results as Dr. Stulberg reported but by replacing 10% glycerol with 10% dimethyl sulfoxide we are rapidly getting up with his magnificant standards.

E. G. D. Murray, Canada - Have any further studies, such as those of Mellanby and Farrell been done? You will recall that they showed that stratified epithelium cultures, in the presence of a great excess of vitamin A, become ciliated cultures and could be completely reversed if the vitamin was removed. It seems to me that you must pay a good deal of attention to the constitution of the medium.

I would like to know whether you find that medium constituents alter the antigenic character of cell lines.

J. F. Morgan, Canada - We have done some studies with vitamin A content of chemically defined media but we have not gone up to extremely high levels. Thus, we have not actually duplicated the effect which Dr. Murray questioned. I don't know of any recent attempts to duplicate the phenomenon.

R. E. Stevenson, USA - Dr. Coons at Cambridge has recently been doing some antigenic analysis on cell lines. He has found that if cells are grown in a medium containing serum which also has the Forsman antigen in it, this does tend to obscure the relationships that he picks up in his mixed agglutination test. Dr. Hall, of the Eli Lilly Co., found, some years ago, that cells grown in serum-containing medium would pick up a sufficient amount of serum to retain some of the antigenic characteristics of that serum after two to three serum-free passages. Thus, this is a possible source of error and confusion which should be avoided.

SUBMITTED PAPERS

ESTIMATION OF STORAGE LIFE OF LIQUID AND DRY STOCK CULTURES OF Pasteurella tularensis AND OF SPORES OF Bacillus anthracis

Ira A. Dearmon, Jr.[1], Michael D. Orlando,
Albert J. Rosenwald, Frederick Klein, Albert L. Fernelius[2],
Ralph E. Lincoln, and Paul Middaugh[3]

A method is presented for estimation of the storage life of liquid and dry bacterial cultures developed to permit rapid evaluation of modifications in preservation techniques and prediction of storage viability. The probit method for interpreting thermal inactivation of bacterial spores published by Fernelius, Wilkes, DeArmon and Lincoln in J. Bact., 73, 300-304, 1958 has been developed in the present paper into a procedure for evaluating studies on storage life of bacterial populations by storage at elevated temperatures with corresponding reduced times for estimation of biological activity.

The quantitative relationships for bacterial cultures during storage for viability, temperature and time was determined for four cultures including liquid and freeze-dried cells of Pasteurella tularensis and liquid and freeze-dried spores of Bacillus anthracis.

Methods for growing, heating and enumerating the liquid suspensions of B. anthracis spores were as described by Fernelius et al. Spore suspensions were freeze-dried by the method of Monk and heated in vacuo in vials and reconstituted for enumeration. The liquid cell suspensions of Pasteurella tularensis were prepared and evaluated by the methods of Hodge and Metcalfe (J.Bact., 75, 258-264, 1958). The dried Pasteurella tularensis are freeze-dried as described by Monk et al. (J.Bact., 72, 368-372, 1956). The cells were suspended in one of three additives before freeze-drying, additive A was a mixture of skim milk and sucrose; additive B a peptone-thiourea-ammonium chloride solution; and C a modification of the Naylor and Smith additive. Liquid suspensions of P. tularensis cells were stored in ampules at 0, 3, 4, 15, 26 and 37 C for periods of 30 minutes to 111 days. The dry cells were held in vacuo in vials stored in desiccators at -18, 3, 4, 15, 27 and 32 and 37 C and sampled at predetermined times.

U.S. Army Chemical Corps, Fort Detrick, Frederick, Maryland
Present addresses: [1] U.S. Army Chemical Corps Operations Research Group, Army Chemical Center, Maryland; [2] National Animal Disease Laboratory, U.S. Department of Agriculture, Ames, Iowa; [3] Grain Processing Corp., Muscatine, Iowa.

Results

Survival of the four cultures was plotted by the method of Fernelius et al. as the probit of survivors against the logarithm of time at the various constant temperatures, including 80, 90 and 100°C for spores of B. anthracis and -18, 3, 15, 27 and 37°C for dried cells of P. tularensis. The survival curves plotted on log-probit paper became nearly linear and the plots for various temperatures yielded parallel lines. Since plots of data have indicated that the probit per cent recovery is linearly related to the logarithm of time for a fixed temperature and is also linearly related to temperature for a fixed time, the joint effect of time and temperature is given by (1) $Y = B_0 + B_1 \log t + B_2$ where Y is probit $(100N/N_0)$ or probit per cent viability; t is storage time; T is storage temperature C; and B_0, B_1, B_2 are the unweighted partial regression coefficients and constants estimable from experimental data. When $(100)(N/N_0)$ is the proportion of viable organisms at time T then equation (1) becomes (2) $t = D \times 10^{-bT}$ where D is a composite constant and $b = B_2/B_1$, the rate of change relating log t and T. When an estimate of b and D are available, estimation of viable storage life, t, can be made directly. Studies with dry P. tularensis will be presented to illustrate the technique.

Dry cells of P. tularensis were stored in vacuo at 5 temperatures and for various time periods within each temperature. The percentage of cells remaining viable at the several storage periods was determined. From these data, we first estimated values of B_1 and B_2 as 1.83 and 0.12, respectively, then, using viable responses for storage at 37 and 27 C derived a tentative B_0 or viability coefficient $B_0 = \bar{Y} + 1.83 \log \bar{t} + 0.12 \bar{T}$, where $\bar{Y}$ is mean probit per cent viability, $\log \bar{t}$, the mean log of storage time, and $\bar{T}$, the mean storage temperature. By using equation (1) with B_1, B_2, and tentative B_0 an extrapolation was made for 80% viability at 3 C which was predicted to be 254 days. The observed life was 279 days. The general relation of shelf-life of dry P. tularensis and storage temperature was estimated to be: $t = 434 \cdot 10^{-0.067T}$, where t is storage time in days, and, T is storage temperature in degrees centigrade. The value of 50 per cent viability was used.

Data collected on cells of dry P. tularensis stored at relatively high temperatures was used to predict storage viability for cells stored at a lower temperature. For the extrapolation to -18 C, the shelf-life was predicted to be 12 years, provided linearity of the function was unaffected by passage through the freezing point of the material. A single observation made after 1 year of storage at -18 C indicated that dry P. tularensis had a viability of 70 per cent. The predicted 50 per cent viability was 98 per cent. Extrapolation far beyond the experimental data should be reserved for estimation of sampling times to test experimental storage conditions.

The accelerated storage technique is useful in experimental work. We determined whether any of the three additives introduced prior to drying would change the storage life of P. tularensis cells. Based on accelerated storage techniques, a decision was made at the end of a 4-day experiment in favor of additive B. Storage at 3 C through 1 year confirmed the decision in favor of additive B as well as the order of superiority of additives C and A. Based on data observed at 37 and 32 C, shelf-life of the P. tularensis cells with the various additives was predicted as follows: B, 584 days; C, 243 days; and A, 148 days. Although the predictions were lower than the values actually observed, they were in the proper order and because they were obtained after only 4 days of observation, tentative decisions could be made that allowed experimental work to proceed.

Heated suspensions of B. anthracis spores yielded viability data similar to those reported previously by Fernelius et al. The predicted 50 per cent viability of spores in liquid of 620 years and for dried spores of 165 years is an extreme extrapolation subject to large prediction errors. The ten suspensions of P. tularensis cells, however, yielded good individual agreement between the observed and predicted values so the mean was used to predict the observed and predicted values with close agreement.

Summary

A method is presented for rapid evaluation of liquid and dry preparations of B. anthracis spores and cells of P. tularensis by storage at elevated temperatures with corresponding reductions in sampling times. The probit per cent viability plotted against the logarithm of time yielded straight lines for each fixed temperature. The lines for the various temperatures were parallel. A prediction equation for each organism yielded estimates for 50 per cent viability at 3 °C which agreed well with observed values for liquid and dry cells of P. tularensis. The predicted values for spores of B. anthracis are almost infinite and can not be verified. The probit equation can be used to evaluate culture drying and storage variables through factorial designs using time and temperature to evaluate the effect of a third variable such as additives, residual moisture, or storage atmosphere.

TECHNOLOGICAL SOLUTIONS TO PROBLEMS IN PRESERVATION AT CRYOGENIC TEMPERATURES

by C. W. Cowley and A. P. Rinfret
Research Laboratory, Linde Company
Tonawanda, U.S.A.

Freezing and thawing are destructive processes if done indiscriminately. The success of low temperature storage is thus critically dependent upon the techniques used in lowering the material to the storage temperature and warming it back to room temperature. The optimum rate of cooling, for instance, varies from one biological material to another. It is important, therefore, that the investigator have techniques available which permit him to study the effect of cooling rates upon the viability of the particular biological he is trying to preserve.

For the purpose of discussion, cooling rates may be arbitrarily divided into four classifications (Table I).

TABLE I

Classification of Cooling Rates

Ultra rapid	150°C/sec	Droplets sprayed onto liquid nitrogen
Rapid	15 to 150°C/sec	Immersion of aluminum containers into liquid nitrogen using insulating coatings
Moderate	1 to 15°C/sec	Immersion of metal or glass containers into liquid nitrogen
Slow	up to 1°C/sec	Controlled

The attainment of ultra-rapid cooling rates by spraying droplets onto liquid nitrogen has been described by Meryman (1) and Rinfret (2,3). The use of insulating coatings to increase cooling rates during immersion in liquid nitrogen was described by Cowley et al. (4). The three highest cooling rates shown in Table I, however, have so far proved to be of value only in the preservation of erythrocytes and some bacteria.

By far the greatest amount of experimentation in the field of low temperature preservation of biological materials is therefore being carried out using "slow" cooling rates as defined in Table I.

Equipment to control the cooling rate of biological samples utilizes one of two general principles. The first involves exposing the sample to an environment which is already at a temperature somewhat below the final value required for the sample. Cooling rate is controlled by insulating the sample container by trial and error. The second principle involves exposing the sample to an environment at essentially room temperature and then lowering the temperature of the environment at the required rate.

But the situation is considerably more complex than this because of the necessity of removing the heat of fusion. Suppose, for
instance, that it is required to cool a sample at the rate of 1°C per
minute. A method of accomplishing this is to put the sample in an
alcohol bath at room temperature and then cool the alcohol at this
rate. Figure 1 shows a cooling curve for such a situation. The
sample cooled at 0.7°C/min up to the heat of fusion and 1.6°C/min
from the completion of the heat of fusion to -50°C. However, immediately after the heat of fusion the sample was actually cooling
at the rate of 4.5°/min. This is merely indicative of the problem
which can arise in the use of this technique. Depending upon conditions we have observed cooling rates immediately after the heat
of fusion of about 10°C/min with the bath temperature cooling at
essentially 1°C/min.

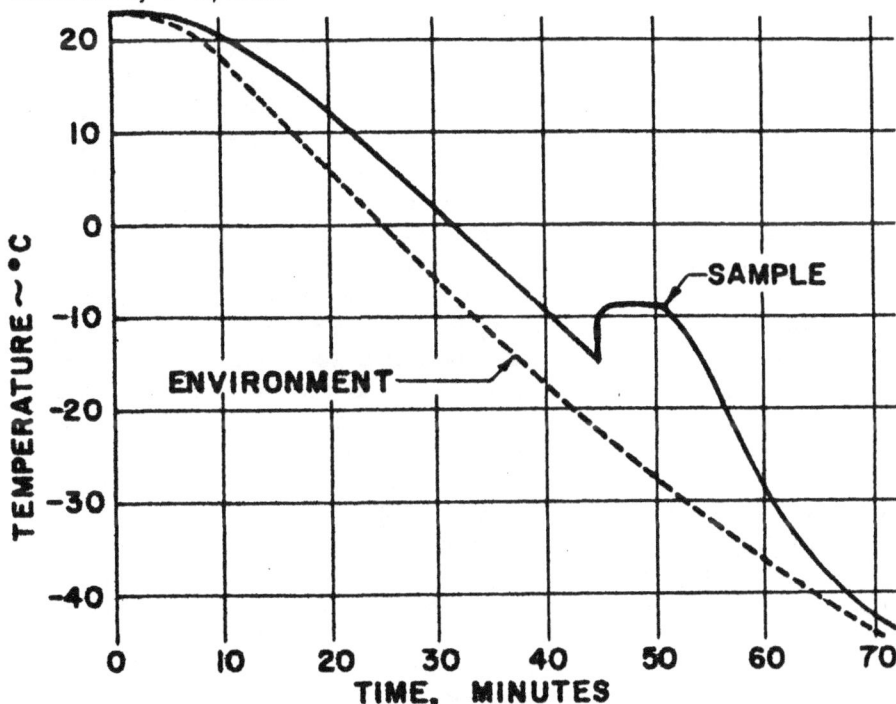

Figure 1
Time-temperature curve for a 5 cc glass vial holding 3 cc
of water. Environment temperature lowered at 1°C/min.

To overcome limitations such as this we use a system in
which the sample itself becomes the reference point. The specimen or specimens are placed in a chamber into which cold nitrogen
gas at approximately -196°C can be injected through solenoid

valves. The chamber is supplied with a fan to provide forced convection and assure uniformity of cooling rates among the specimens. One of the specimens is used as the control reference or, if this is not desired due to sterility requirements, a "dummy" sample of identical thermal characteristics can be used. One leg of a differential thermocouple is placed in the control sample while the other leg remains outside. A conventional thermocouple is also included in the control sample to continuously monitor the temperature.

The rate at which the specimens will be cooled is determined by the magnitude of the output signal from the differential thermocouple. This signal is fed to a relay amplifier which has been biased by an adjustable reference circuit. The higher the bias setting the greater will be the temperature difference maintained between the sample and its environment and the higher will be the cooling rate.

Using this system of control we are able to cool samples at rates ranging from about 0.5°C/min up to 20°C/min. Cooling rates in the liquid and solid phase can be changed simply by an appropriate adjustment of the reference signal. One of the most important features, however, is the fact that the time spent by the samples in the heat of fusion can also be controlled. This is proving to be an extremely important variable in optimizing cooling conditions for some biological materials.

There are some limitations to this control system. In common with most materials, the specific heat of biologicals is a function of temperature. This means that some change in the differential temperature setting is needed to maintain an absolutely constant cooling rate over a wide temperature range. Since the cooling rate for most biologicals, however, is most critical over the relatively small temperature range from freezing to -50°C this does not appear to be a severe problem.

The rate at which a sample cools will depend not only upon the imposed temperature driving force, but upon its thermal characteristics, cross section and the geometry of the container into which it has been placed. This means that a new calibration of the reference setting versus cooling rate is required whenever any of these variables are changed. In actuality this is a much smaller problem than it would appear to be. Our investigators are able to obtain desired rates quickly and reproducibly for a wide variety of samples, sample cross sections and geometry. Figure 2 shows a picture of the complete unit including a temperature recorder and the liquid nitrogen cylinder which is used as the source of refrigeration.

Figure 3 illustrates typical cooling curves obtained using the differential temperature control system.

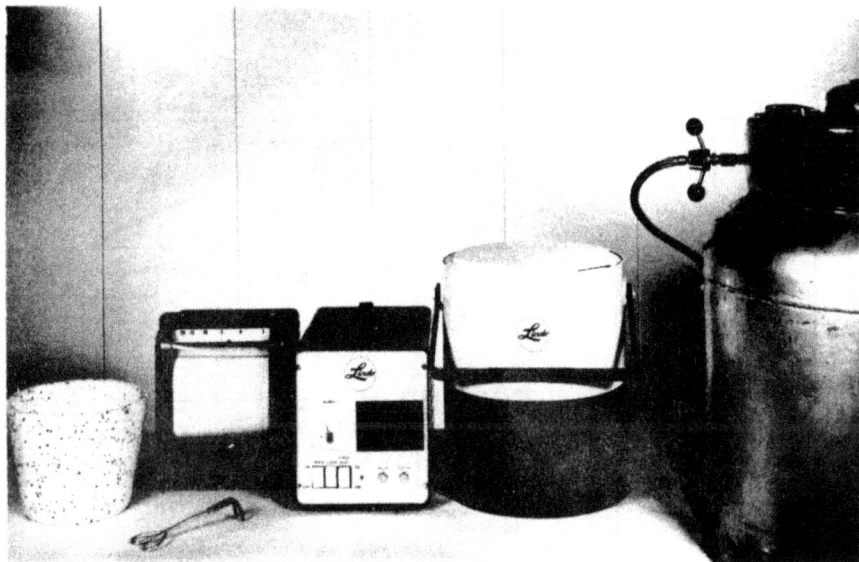

Figure 2

Equipment for controlled rate freezing of biological materials

Controlled rate cooling equipment such as I have described is in routine use at our laboratories in studies of techniques for the low temperature preservation of tissue cultures, bone marrow, semen and micro-organisms. We have been pleased with its operational dependability and reproducibility.

In order to obtain even greater ease of operation and higher precision, however, we have designed specific circuitry which will allow complete pre-programming of the cooling operation. This system will include automatic variation of rates in different temperature regions and pre-selection of the time to be spent in the heat of fusion.

References
1. Meryman, H.T. and Kafig, E. Proc. Soc. Exp. Biol. and Med. 90: 587. 1955.
2. Rinfret, A.P. Ann. N.Y. Acad. Sci. 85: 576. 1960.
3. Rinfret, A.P. and Doebbler, G.F. Biodynamica 8: 181. Nov. 1960.
4. Cowley, C.W., Sawdye, J.A. and Timson, W.J. Biodynamica 8: 317. Dec. 1961.

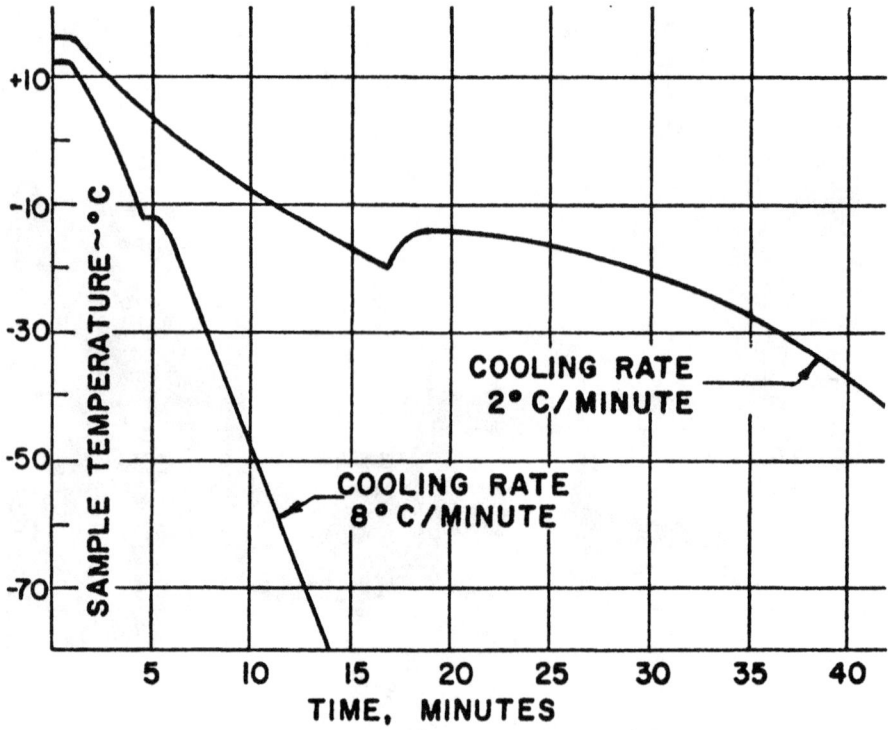

Figure 3

Typical cooling curves using differential temperature control

FUNDAMENTALS IN THE APPLICATION OF CRYOGENIC TEMPERATURES TO THE MAINTENANCE OF VIABILITY IN MICRO-ORGANISMS

by S. W. Moline, A. W. Rowe, G. F. Doebbler and A. P. Rinfret
Research Laboratory, Linde Company
Tonawanda, U.S.A.

Preservation of bacteria, viruses, tissue cultures, bone marrow, blood, and other biological materials at sub-zero temperatures has become a common practice among biologists in a wide variety of scientific pursuits. Recent advances in the techniques of low temperature preservation have resulted in the successful storage of many tissues or organisms previously considered to be destroyed during cooling, storage, or subsequent warming processes.

The objectives of long-term preservation include retention of viability, morphology, growth characteristics, chromosomal pattern, and enzyme activity. My colleage, Mr. Cowley, has classified cooling regimes into four categories and has described equipment that will enable a worker to cool samples at controlled slow rates. In addition, this apparatus has permitted us to study the effect of still another important variable in any freeze-thaw process - the length of time the specimen spends in the temperature region in which the latent heat of fusion is removed.

Bone marrow specimens are cooled with the greatest retention of viability, whether measured by amino acid incorporation or cellular respiration, if slow cooling and rapid removal of the heat of fusion are employed (1). This is illustrated in the first table.

TABLE I
Effect of the time required to remove the latent heat of fusion on the viability of bone marrow samples cooled at a rate of 1°C per minute

Latent heat of fusion period	Viability as per cent of control	
	Glycine incorporation (15% DMSO)	Respiration (15% Glycerol)
1-2 minutes	52	45
8-10 minutes	35	33
15-25 minutes	2	3

However, within the definition of "slow cooling", precise rates may also affect post-thaw recovery of the tissue or organism. With bone marrow, optimum results are obtained with a cooling rate of 1°C/min once the heat of fusion has been removed. See Table II. Other cells have a much lower sensitivity to slow cooling rates. Kite and Doebbler (2) have shown that cooling rates

TABLE II

Effect of cooling rate on the viability of bone marrow
samples once the heat of fusion has been removed

Cooling rate	Viability as per cent of control respiration
1°C/min	78
3-4°C/min	56
8-9°C/min	45

between 1° and 45°C/min gave comparable cell recoveries with
HeLa and L-strain mouse fibroblasts. Mazur (3) found that optimum survival of yeast exposed to sub-zero temperatures depended
upon slow cooling and rapid warming. Fungi, protozoa, algae,
rusts, bacteriophages have also been shown to survive a freeze-
thaw regime if cooled slowly with a protective additive, stored at a
sufficiently low temperature, and warmed rapidly (4).

Bacteria, however, are less sensitive to cooling and thawing
rates than are the tissues and organisms previously mentioned.
Numerous reports over the past 75 years have described bacterial
survival following exposure to low temperatures, usually in qualitative terms (5). We have examined the effects of ultra-rapid cooling on the quantitative survival of five organisms: Azotobacter
vinelandii, Escherichia coli, Staphylococcus aureus, Saccharomyces
cerevisiae, and mycelial suspensions of Aspergillus niger. The
cell suspensions or mycelial particles were sprayed as droplets
onto a moving surface of sterile liquid nitrogen. The frozen droplets were collected and thawed rapidly by immersion in a 0.15 M
NaCl solution at 37°C. Cooling and warming rates of several hundred degrees per second have been calculated for these conditions
(6). The results of this experiment are seen in Table III.

TABLE III

Survival of various micro-organisms following droplet
cooling in liquid nitrogen and rapid warming

Organism	Number of cells frozen	Per cent Survival
Azotobacter vinelandii	9×10^7	28 ± 8
Escherichia coli	9×10^8	100 ± 4
Escherichia coli Strain B	--	not given (4)
Aerobacter aerogenes	--	100 (7)
Staphylococcus aureus	1×10^9	91 ± 11
Staphylococcus epidermidis	--	60
Serratia marcescens	--	not given (4)
Aspergillus niger	1×10^3	22 ± 1
Saccharomyces cereviscae	9×10^9	42 ± 7

It is seen that the survival of two of the bacterial strains is
extremely high while one strain is significantly destroyed. The
bacterium Serratia marcescens was also shown to survive this
freeze-thaw procedure by Clark (4). The yeast and fungus examined in our study exhibited definite but low survival.

Thus we have seen that various cells and organisms exhibit
a wide variety of response to cooling and warming treatments.
Some are extremely sensitive and require rather precise control
whereas others have a high tolerance to various freeze-thaw procedures. Before generalizations can be drawn with respect to differences between bacterial species, more detailed studies of the
biophysical and biochemical parameters involved in the freezing
of micro-organisms must be made.

The last factor that will be discussed here is concerned with
the temperature required for long-term survival of micro-
organisms and cells.

Refrigeration between 2° and 10°C on agar slant or in broth
cultures has been used to preserve bacteria, fungi, viruses, and
protozoa. Subculturing once or twice a year is necessary. Cells,
may be preserved only for very short periods of time at these
temperatures.

A second method used in preservation has been commonly
referred to as "deep-freezing". Using either a protective additive
or medium and cooling at controlled or uncontrolled rates, many
bacteria, viruses, and protozoa have been preserved. Cells should
be cooled under controlled conditions prior to storage. This
method, although generally quite convenient, is fraught with danger.
Many bacteria and viruses may be preserved for long periods of
time at temperatures above -78°C, while others are gradually
affected by this treatment. Viral titers are lost and post-thaw recoveries of most of the samples diminish. Certainly storage at the
lowest temperature in this range is preferred.

The third procedure has been termed ultra-low or cryogenic
storage. Although an acceptable upper temperature limit for long-
term preservation has not been definitely established, it is believed
that the storage temperature should be somewhere below -100°C.
Ice crystal modification or growth, which is considered to be one
cause of cell destruction, has been found to occur at temperatures
above -100°C. Studies with bull semen and red blood cells stored
at various low temperatures have shown retention of viability after
storage below -100°C. This is seen for red blood cells in Figure 1
(8).

Although many cells and cell lines have been stored at -78°C
without significant loss of viability, often storage time was of
comparatively short duration. Other studies have definitely established loss of viability after storage at -78°C for one year whereas

FIGURE 1

Red Cell recovery after storage in the frozen state

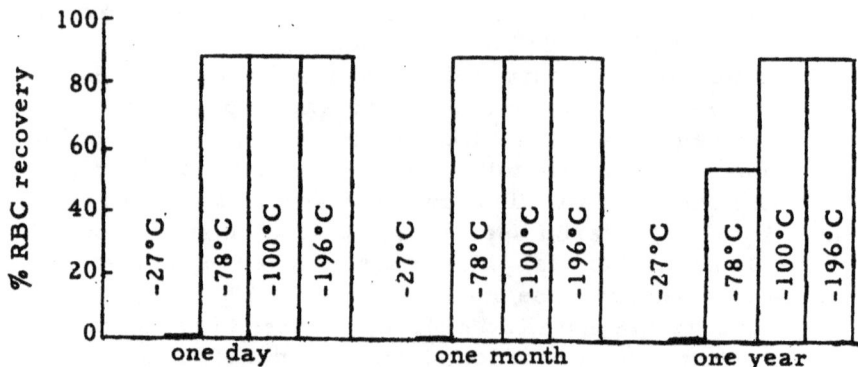

storage at -196°C was shown to retain the initial viability. For this reason, the American Type Culture Collection and the Cell Culture Collection Committee both advocate storage at -170°C or lower to assure complete retention of activity. Until further data on the effect of prolonged storage of specimens at various temperatures has been obtained, it is recommended that cryogenic storage temperatures be used.

References

1. Rowe, A.W. and Rinfret, A.P. Blood. in press.
2. Kite Jr., J.H. and Doebbler, G.F. Fed. Proc. 20:149. 1961.
3. Mazur, P. Ann. N.Y. Acad. Sci. 85:610. 1960.
4. Clark, W.A. Proc. Low Temp. Microbiol. Symp. p.291, Campbell Soup Company, New Jersey. 1961.
5. Smith, A.U. Biological Effects of Freezing and Supercooling. pp.74-137. Williams and Wilkens, Baltimore. 1961.
6. Rinfret, A.P. Ann. N.Y. Acad. Sci. 85:576. 1960.
7. Glycerine Facts, Summer 1962, No.2.
8. Rinfret, A.P. Fed. Proc. Symp. and Special Reports. in press.

PRESERVATION OF PARASITIC PROTOZOA IN LIQUID NITROGEN

by Louis S. Diamond
Laboratory of Parasitic Diseases,
National Institutes of Health, Public Health Service
and
Harold T. Meryman and Emanuel Kafig
Naval Medical Research Institute,
National Naval Medical Center
Bethesda, U.S.A.

Introduction

The use of cryogenic techniques for the preservation of trophic stages of parasitic protozoa is not new; Entamoeba, Trichomonas, Plasmodium, Toxoplasma, and Trypanosoma have been successfully preserved in this manner (6). However, in each instance, the temperature of storage has been at or near dry-ice (-79 C). At such a relatively high temperature the limits of preservation of living cells can be reckoned in months, at most a few years. These limits are set, to a great extent but not solely, by the fact that at this temperature ice crystals, which in one way or another are so intimately concerned with the causation of death in the frozen state, can form and grow (5). Since ice crystal formation and growth can not occur below the recrystallization point of water (-130 C), theory suggests the use of storage temperatures below this point for long term preservation, of the order of several years. The use of such temperatures is additionally suggested by the fact that degradative biochemical reactions, also responsible for death in the frozen state, are greatly slowed below this point.

In recent years, liquid nitrogen has come into use as a refrigerant for long term storage. It is ideally suited for this purpose. It is chemically inert, it vaporizes without residue and provides a storage temperature (-196 C) at which little, if any, chemical or physical change can occur.

The investigations reported here were undertaken to determine the feasibility of using liquid nitrogen for the preservation of parasitic protozoa. The details of the techniques used and a summary of the data obtained from our storage records are presented. A portion of this work has been published (2).

Materials and Methods

Entamoeba histolytica (Strain 200:NIH), Trichomonas foetus (BP-4:Beltsville), T. gallinae (DP-3 = Jones' Barn strain), T. hominis (HuF-2:Beltsville), T. vaginalis (C$_1$:NIH), Trypanosoma cruzi (2380-260:PRR) and T. ranarum (13B) were used in these studies.

The trichomonads and trypanosomes were obtained from axenic cultures (1,3); the amoebae from monoxenic cultures grown in association with a <u>Crithidia</u> sp. (7). Forty-eight to 72-hour growths of trichomonads, 72-hour growths of amoebae and 7-day growths of trypanosomes were used.

Organisms to be preserved were concentrated by centrifugation, and suspended in fresh medium. In the case of the trichomonads, which were grown in medium containing 0.05% agar (1), agar-free medium was used as the suspending agent. To these suspensions, dimethylsulfoxide (J. T. Baker Chemical Co.) in a final concentration of 5% was added as a protective agent (4). After this, the trichomonad and amoebae-<u>Crithidia</u> suspensions were allowed to equilibrate 15 and 30 minutes, respectively, at 35 C; the trypanosomes for 15 minutes at room temperature. They were then distributed in 0.2 ml portions to screw-capped vials (15 x 45 mm).

Each vial of trypanosomes contained 1/15 the yield from a 7-day culture; the actual number of organisms was not determined. On the other hand, each vial of trichomonads contained either 900,000 or 2.5 million organisms; the amoebae-<u>Crithidia</u> vials contained 290,000 or 780,000 amoebae.

A two- or three-step freeze-cycle was used. The vials were placed in a Canalco Slow-Freeze unit set at an initial temperature of 0 C. After allowing a few minutes for temperature equilibration the specimens were cooled to -35 C at a rate of approximately 1°/min. At this point, they were removed from the Canalco unit and handled as follows: The amoebae were transferred to a dry-ice cabinet for 1 to 96 hours, then placed in liquid nitrogen, or they were plunged directly into liquid nitrogen; the trichomonads and trypanosomes were plunged directly into the nitrogen.

The frozen specimens were stored in a liquid nitrogen refrigerator (Linde Co., Model #LNR-25-B) either in the liquid or its vapors.

To thaw specimens, the vials were momentarily immersed and rapidly swirled about in a 45 C watherbath containing a few drops of 25%-Aerosol O. T. (American Cyanamid Co.). The Aerosol O. T. was added to improve heat transfer. Immediately following thawing, fresh medium was added to each vial; 3 ml being added to the amoebae and trichomonad suspensions, 1.5 ml to the trypanosomes. The vials were then placed in an incubator of the appropriate temperature. After one hour, they were removed, placed under the low power objective (10x) of a compound microscope and their contents examined for motile organisms. If this procedure failed to reveal organisms, a drop of fluid was removed from the bottom of the vial and examined as a slide preparation. Finally, the samples were reincubated. Since trypanosomes as a rule do not grow well in liquid medium, the entire sample was transferred onto a blood agar base for incubation (3).

The criterion of successful preservation was based on the ability of the protozoa to reproduce in culture. As a further test,

at least one sample from each experimental batch was twice transferred serially. To insure growth in the case of the amoebae, it
was necessary to add additional Crithidia to the suspension either
just prior to freezing or immediately following thawing.

Results and Discussion

    In evaluating the efficiency of the technique, samples were
thawed and cultured routinely 24 hours after freezing. A visual
estimate was made of the number of motile organisms present
1 hour after thawing, then daily during the period of incubation. In
this way a base line of observations for comparison with the specimens stored over extended periods of time was made.

    To date, cultures of Entamoeba histolytica have been obtained
from samples stored as long as 14 months. Yields obtained from
samples subjected to transient storage in dry-ice prior to storage
in liquid nitrogen were no better than those stored continuously in
liquid nitrogen following initial cooling to -35 C. For this reason
transient storage in dry-ice was subsequently eliminated from the
procedure.

    Viable cultures have been obtained from samples of Tricho-
monas gallinae, and T. vaginalis stored 10 months and T. foetus
and T. hominis stored 5 months. Trypanosoma cruzi and T. ranarum have so far survived storage up to 4 months.

    Examination of the samples one hour after thawing also per-
mitted an evaluation of the efficiency of the technique insofar as
the freeze-thaw cycle itself was concerned. Based on a visual estimate of the number of motile organisms observed after thawing,
trypanosomes were found to survive freezing and thawing best, the
amoebae least.

    No difference in yields were found between samples of a
given species thawed 24 hours after freezing and those thawed
after the longest period of storage. This indicated absence of decay
during storage, a common occurrence at dry-ice temperature.
Moreover, no sample tested failed to produce viable cultures. In
the case of E. histolytica, cultures were obtained from each of 110
samples stored anywhere from 1 day to 14 months.

    Screw-capped vials were favored over the standard hermati-
cally sealable ampoules because of the former's usefulness as a
culture vessel following thawing of the samples and the potential
explosion hazard the latter presented during thawing if not adequately sealed.

    In the early experiments storage was accomplished by im-
mersion in liquid nitrogen. When it was found that the rubber liner
of the cap did not always offer an effective seal against the liquid
nitrogen, storage was continued in the vapors of the nitrogen to
avoid the possibility of: 1) microbial contamination from this source,
2) an explosion during thawing. Storage in the vapors is now routine.

Summary

Trophic stages of a selected group of parasitic protozoa have been recovered in a viable state after subjection to the temperature of liquid nitrogen (-196 C). To date, these protozoa have survived storage at this temperature without noticeable decay other than that which occurs during the freeze-thaw cycle for 14 months in the case of Entamoebae histolytica; 10 months in the case of Trichomonas gallinae and T. vaginalis; 5 months for T. foetus and T. hominis and 4 months for Trypanosoma cruzi and T. ranarum.

References

1. Diamond, L. S. The establishment of various trichomonads of animals and man in axenic cultures. J. Parasitol. 43: 488-490. 1957.

2. Diamond, L. S,, Meryman, H. T. and Kafig, E. Storage of frozen Entamoeba histolytica in liquid nitrogen. J. Parasitol. 47(Suppl.): 28. 1961.

3. Diamond, L. S. and Rubin, R. Experimental infection of certain farm mammals with a North American strain of Trypanosoma cruzi from the raccoon. Exper. Parasitol. 7: 383-390. 1958.

4. Lovelock, J. E. and Bishop, N. H. W. Prevention of freezing damage to living cells by dimethylsulfoxide. Nature 183: 1394-1395. 1959.

5. Meryman, H. T. General principles of freezing and freezing injury in cellular materials. In Freezing and drying of biological materials. Edited by H. T. Meryman. Ann. New York Acad. Sci. 85: 503-509. 1960.

6. Smith, A. U. Effects of freezing and thawing on microorganisms. pp. 74-137. In Biological effects of freezing and supercooling. Edward Arnold Ltd., London. 1961.

7. The techniques of cultivating the amoebae with the Crithidia sp. will be published.

THE EFFECT OF SOME TREATMENTS FOR STARVATION UPON THE VIABILITY OF YEAST CELLS THROUGH LYOPHILIZATION

by G. Terui and J. Ikeda
Osaka University
Osaka, Japan

Introduction

The viability and the change in characteristics of microbes through lyophilization and storage are dependent, firstly, upon the external conditions affecting the cells during the treatment and, secondly, upon the culture conditions under which the cells have been made up. Despite some information (1) dealing with the lyophilization of yeasts, little is available on the second problem. We have observed that nitrogen-starved yeast cells are generally more resistant not only to heat but also to freeze-drying (2). The present study was intended to determine the effect of vitamin- and nitrogen-starvation upon the viability of yeasts through lyophilization and storage. In addition, the change in bios-pattern was investigated.

Methods and Materials

Organisms – The results with two strains of Saccharomyces cerevisiae, 2 and 209, were listed. Other than these, 15 species of Saccharomyces selected from our culture collection were employed in a few instances.

Cultivation and growth medium – Except in one of our experiments, in which J. White's medium was used, Burkholder's medium was adopted as the basal growth medium. Cultivation was at 28°C for 36 hours on a shaker.

Starvation culture – The cells harvested from the growth medium were washed twice with water and suspended in a starvation medium having a volume corresponding to the original culture. The suspension was incubated on a shaker at 28°C for 48 hours unless otherwise indicated. Starvation medium had essentially the composition of Burkholder's except that a factor, or factors, was omitted. The vitamins here concerned were more or less stimulatory to growth; none was required absolutely for growth of the above two strains.

Suspending medium – Phosphate buffer (M/15; pH 7) was used throughout as the previous experiments showed that this was as good a suspending medium as horse serum or skimmilk.

Lyophilization – The cells harvested from a culture were washed twice with water and re-suspended in phosphate buffer. The suspension was dispensed in 6 ml ampoules, 0.5 ml suspension or 5-7 mg dry cell weight per ampoule. Prior to evacuation, prefreezing was conducted slowly in a cooling tray. Samples were dried at manifolds or on the perforated holder in a chamber for

7-8 hours. During the treatment, temperature was increased until it approached 30°C. The moisture content of lyophilized samples, determined by drying at 105°C or by K. Fischer's method, was 2 per cent or a little lower.

Storage – Ampoules sealed at the existing vacuum were stored at ca. 15°C for 6 months unless otherwise indicated.

Viable counts – A sample of lyophilized culture was added to a definite volume of sterile water at room temperature. From the cell suspension, a series of ten-fold dilution was made. Viability was determined by counting colonies that developed on Burkholder agar plates after incubation at 30°C for 48 or 72 hours. Viability before lyophilization was determined in a similar manner.

Fermentability – Durham tubes were used for determining fermentable sugars according to Lodder's method.

Results and Discussion

The effectiveness of pre-treatment for nitrogen-starvation in maintaining viability through lyophilization and storage has been proved by us with a variety of yeasts whose growth is stimulated remarkably by pantothenic acid. An example is shown in Table I.

TABLE I

Effect of nitrogen-starvation (S. cerevisiae, strain 2)

Pre-treatment	Growth medium	Survival (%)			
		1	3	5	7 (months)
None	J. W.	49	31	21	11
	Enriched* J. W.	81	63	53	43
N-starvation	J. W.	72	55	44	35
(for 24 hrs)	Enriched* J. W.	104	87	76	67

* Enriched with biotin and aspartic acid, resp., to 10-fold in concentrations.

Enrichment of growth medium with biotin and aspartic acid was shown to be effective in obtaining cells resistant to the lyophilization treatment. The effect of nitrogen-starvation, however, was obscure or rather negative in some strains which do not require pantothenic acid for good growth, as shown later by an example. Then the vitamin-starvation treatment was conducted in the presence or absence of nitrogen source. The cells of S. cerevisiae, strain 2, which is auxotrophic to pantothenic acid to a great extent, became highly susceptible to lyophilization and storage when sub-cultured in a pantothenic acid-deficient.medium in the presence of a nitrogen source: almost all the cells died immediately after lyophilization. Pyridoxine-starvation culture also resulted similarly, though pyridoxine was nearly nonessential to this strain. Starvation culture with regard both to pyridoxine and pantothenic acid gave relatively high survival rates after lyophilization as illustrated in Fig. 1(a).

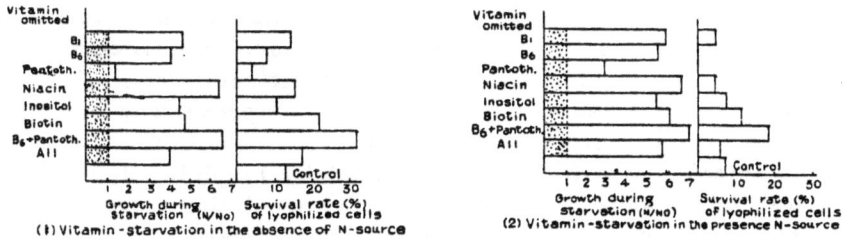

Fig. 1(a). Effect of vitamin-starvation (S. cerevisiae, strain 2)

A unique theory for explaining the above results is difficult to deduce. Nitrogen-starvation culture in all instances was shown to be effective whether vitamin is omitted or not.

S. cerevisiae, strain 209, which requires pyridoxine and inositol for good growth but is almost independent of pantothenic acid, produced cells which became, by nitrogen-starvation, rather susceptible to lyophilization as shown in Fig. 1(b).

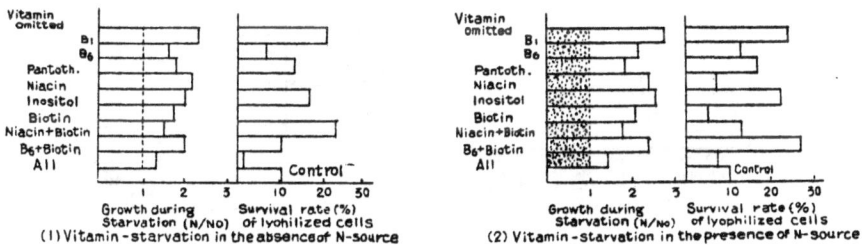

Fig. 1(b). Effect of vitamin-starvation (S. cerevisiae, strain 209)

In this case, starvation treatment with regard to a required vitamin did not cause the decrease in survival through lyophilization but brought about rather good effect for maintenance. A medium deficient in biotin or nicotinic acid, produced highly susceptible cells when used for the subculture of this strain (Fig. 1 b). Though there were some exceptions, nitrogen-starvation might be regarded as an effective means to enhance the survival after lyophilization, and this seems to be in accord with the fact that the

majority of strains belonging to <u>Saccharomyces</u> requires pantothenic acid for normal growth. Previous investigations on the
variation of fermentability and assimilability during lyophilization
and storage have shown that these characteristics are highly
stable but in certain instances galactose and raffinose fermen-
tabilities are lost by the treatment, e.g. galactose fermentability
of <u>S</u>. <u>logos</u>, strain 278, and raffinose fermentability of <u>S</u>. <u>carti-</u>
<u>laginosus</u>. Nitrogen-starvation was tried for preventing such a
variation but the results were negative.

     Variation of vitamin requirement by lyophilization and
storage, so far as it was observed at the population level, scarcely
took place where reasonable survival rates were maintained.
Some of the lyophilized cells failed to grow in a vitamin-deficient
medium in which original untreated cells could grow. In order
to know whether the loss of vitamin-synthesizing ability observed
at the cellular level is a transient one, colonies grown from
lyophilized cells on the complete agar medium were transferred
to a vitamin-deficient agar plate, in order to obtain replica plate
culture. A representative example is shown in Fig. 2 which indicates that most of the observed losses in vitamin-synthesizing
abilities of lyophilized cells are transitory in nature detectable
only at the cellular level.

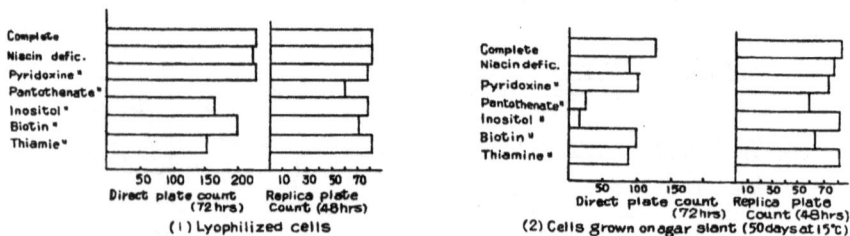

Fig. 2(a). Comparison of bios-pattern of lyophilized culture with
that of old agar culture (<u>S</u>. <u>cerevisiae</u>, strain 2)

Summary

     Nitrogen-starvation pre-treatment is generally effective in
increasing survival after lyophilization and storage of Saccharomyces yeasts. But with strains highly auxotrophic to pantothenic
acid, the effect was obscure or contrary. Such a pre-treatment
was ineffective in preventing rarely observed loss of fermentability. The effect of vitamin-starvation treatment was sometimes

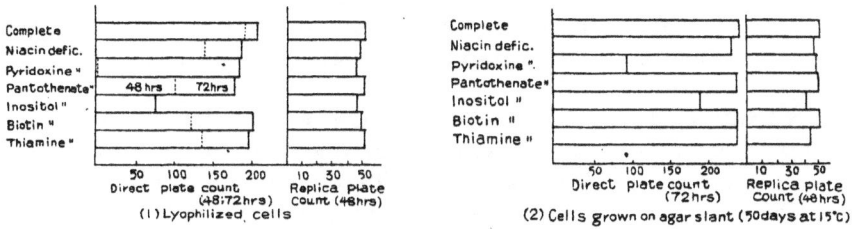

Fig. 2(b). Comparison of bios-pattern of lyophilized culture with that of old agar culture (S. cerevisiae, strain 209)

remarkable either positively or negatively. The general relationship between requiring vitamins and the effect of vitamin-starvation treatment could not be deduced. Transient loss of vitamin-synthesizing ability was observed to occur very frequently in lyophilized cells.

References

1. Wickerham, L. J. et al. Wallerstein Lab. Comm. 5: 165. 1942.
    Atkin, L. et al. ibid. 12: 365. 1949.
    Kirsop, B. H. J. Inst. Brew. 61: 466. 1955.
2. Terui, G. and Ikeda, J. presented at annual meetings of Soc. Ferment. Technol. (Jap.) and Agric. Chem. Soc. Jap. 1958-1962.

QUANTITATIVE STUDY OF DAMAGE AND SURVIVAL OF BACTERIA DURING FREEZE-DRYING AND DURING STORAGE OVER A TEN YEAR PERIOD

by Arthur P. Harrison, Jr. and Michael J. Pelczar, Jr.
Vanderbilt University, Nashville, U.S.A.
University of Maryland, College Park, U.S.A.

A broth culture was centrifuged, the pellet taken up in double-strength skim milk, and this suspension dispensed in 0.2 ml portions into ampoules (10 mm x 35 mm). The ampoules were frozen in an ethanol-dry ice bath, were placed in a jar at room temperature connected to a vacuum pump assembly, and the frozen suspensions were dried at 70-100 micra Hg overnight. The ampoules were removed from the vacuum jar, were placed in phials (14 mm x 60 mm), and the latter sealed at 70-100 micra. Assays were carried out prior to the freezing, immediately after freezing (thawed at room temp.), immediately after the drying period, and after various intervals of storage at 8°C. The phials were opened, the ampoule was wiped with ethanol, was air dried, and then was pulverized by shaking within a bottle containing saline and several glass marbles. From this initial dilution further serial dilutions were prepared and aliquots were plated on suitable agar. Table I summarizes viability data with 12 species.

TABLE I

Viability after freezing, drying, and freeze-dried storage
(All assays represent the number of cells in the ampoule)

Species	Initial count	After freez.	After dry.	After storage for			
				1 wk.	8 wks.	12 wks.	10 yrs.
E. coli	9×10^9	9×10^9	8×10^9	3×10^8	2×10^8	2×10^8	2×10^8
V. costi.	3×10^8	4×10^8	----	5×10^4	5×10^4	2×10^4	2×10^1
L. bifidus	4×10^8	3×10^8	3×10^8	1×10^9	4×10^8	3×10^8	2×10^7
L. fermenti	7×10^8	6×10^8	3×10^8	2×10^8	1×10^8	1×10^8	1×10^8
Ac. melano.	7×10^8	7×10^8	5×10^7	3×10^7	8×10^7	----	2×10^1
Aer. aerog.	5×10^9	5×10^9	2×10^9	9×10^9	1×10^9	9×10^8	7×10^8
Ps. species	7×10^9	9×10^9	1×10^9	7×10^8	4×10^8	9×10^8	4×10^8
Ps. aerugin.	4×10^9	3×10^9	1×10^8	5×10^7	6×10^7	----	7×10^5
Ps. chloro.	1×10^{10}	9×10^9	8×10^6	----	1×10^5	6×10^5	3×10^5
Prot. vul.	3×10^9	3×10^9	6×10^7	5×10^7	1×10^7	----	5×10^6
Ser. marces.	1×10^9	9×10^8	2×10^8	2×10^8	1×10^8	1×10^7	5×10^7
M. albus	5×10^9	5×10^9	3×10^9	2×10^9	7×10^8	7×10^8	9×10^8

Freezing (with thawing) causes little, if any, loss of titer. But the subsequent manipulations cause a measurable loss of titer which varies among species. The loss of viability is either due to the drying process itself or to the exposure of the dried material to the air during the brief interval between the 2 evacuations (1). Semilog-time plots predict the longevity of the dried suspensions and demonstrate differences in sensitivity among species. Extrapolation from the 12 week and 10 year points indicates that the ampoules containing Vibrio and Acetobacter will become sterile within several decades, whereas most ampoules (assuming constant death rate) will contain viable cells after centuries. (Of course longevity will depend upon the number of cells initially present.) Sensitivities may be compared by means of death rates. Thus, the number of years required to cause a mere ten-fold drop in titer are 100 for Escherichia coli, 3 for Vibrio costicolus, 9 for Lactobacillus bifidus, 40 for L. fermenti, 1 for Acetobacter melanogenum, 40 for Aerobacter aerogenes, 25 for a psychrophilic Pseudomonas, 5 for Ps. aeruginosa, 25 for Ps. chlororaphis, 35 for Proteus vulgaris, 17 for Serratia marcescens, and many 100's for Micrococcus albus. The Pseudomonadaceae are among the most sensitive. Appreciable variation in sensitivity occurs among the Enterobacteriaceae. All bacteria appeared unaltered in appearance and general growth characteristics, but quantitative physiological tests were not undertaken.

An alternate method of lyophilization was as follows. Growth from an agar slant culture was suspended in the skim milk. Aliquots (0.2 ml) were placed in ampoules (7 mm diam. x 80 mm tubes with 11 mm diam. bulbs at the distal end). Freezing and drying was the same as above, except that the ampoules remained in the freezing bath during drying and were sealed while attached to the manifold. This method is easier and avoids exposure of the dried material to the air. Storage was at room temperature. The higher temperature should increase the death rate (2, 3, 4, 5), accordingly, might increase selection of genotypes within the population. Viability tests were only qualitative, but detailed comparisons of physiological traits before and after lyophilization were carried out. Species of Streptococcus, Betacoccus, Pediococcus, and Lactobacillus, including the anaerobic L. bifidus, stored 5 years showed no change in mass culture characters, whereas some Bacteroides sustained alterations in their fermentation spectrum (table II). However, the products from glucose fermentation were unaltered: lactic acid remained the principle end-product and the zinc lactate remained the levo-rotating isomer. Morphology, temperature requirements, and gross growth characteristics also remained unaltered. After 3 additional years all ampoules had become sterile. Thus it is likely that great loss of titer had occurred at the time of the tests (table II). A pleomorphic Bacteroides, strain 29-9, did not survive even the 5 year period. Thus, some gram-negative, intestinal anaerobes are much more sensitive to lyophilization than are the Lactobacteriaceae and other bacteria.

TABLE II

Final pH values of broth cultures of 2 Bacteroides strains
before and after lyophilization and storage for 5 years
(Significant fermentation has been italicized)

| Substrate | Strain 21-28 | | Strain 28-30 | |
	Before	After	Before	After
Glycerol	6.9	6.6	7.0	*4.7*
Xylose	*4.8*	*5.1*	*4.7*	6.4
Arabinose	*4.8*	6.5	*4.7*	6.4
Sorbitol	*5.3*	6.7	*5.1*	6.4
Mannitol	*5.5*	6.6	*5.3*	6.5
Glucose	*4.9*	*5.1*	*4.9*	*4.8*
Galactose	*5.0*	*4.8*	*4.8*	6.2
Trehalose	*4.8*	6.6	*4.8*	6.5
Maltose	*4.8*	*5.0*	*4.9*	6.6
Cellobiose	6.8	6.6	*4.8*	6.4
Melibiose	*4.8*	6.5	6.5	6.5
Lactose	*4.9*	6.5	*4.7*	6.2
Raffinose	*4.8*	6.6	*4.8*	6.8
Salicin	*5.0*	6.6	6.7	6.7
No substrate	7.0	6.7	7.0	6.6

Frozen storage as pellets is suitable for some bacteria that
are sensitive to lyophilization. A broth culture is centrifuged, the
broth decanted, and the centrifuge tube with pellet placed directly
in the deep-freeze unit. Quantitative experiments with this method
have been undertaken only over a 12 week period (table III), but it
is noteworthy that survivals may be better than with lyophilized
material (table I). Storage as thick water suspensions may be as
good, or better, since many bacteria are not affected adversely
thereby. For example, it has been observed repeatedly that E. coli
titers are not diminished as a result of freezing and thawing in
distilled water (6, 7, 1).

Freezing agar slant and butt cultures at -22° gives good results and is convenient. It is important that toxic end-products of
growth be minimized, as by using a glucose concentration of 0.25%
or less. The following bacteria have been maintained 2-3 years
with no detectable change in morphology, growth characteristics,
or physiological traits: 10 species of Streptococcus, 7 species of
Bacillus, 6 species of Lactobacillus, 3 species of Acetobacter, 2
species of Microbacterium, 2 species of Propionibacterium, 2
species of Sarcina, 3 species of Betacoccus, 2 species of Micrococcus, Chromobacterium, Pediococcus, Proteus, Escherichia,
Aerobacter, Alkaligenes, Pseudomonas, and Saccharomyces.

Anaerobic Bacteroides and Lactobacillus have been stored aerobically for a year with no change in characters noted.

TABLE III
Viability of bacteria after freezing as pellets

Species	Temp. (°C)	Initial count	1 wk.	2 wks.	4 wks.	8 wks.	12 wks.
E. coli	-25	1×10^9	1×10^9	1×10^9	9×10^8	----	1×10^9
	-60	1×10^9	9×10^8	1×10^9	1×10^9	----	9×10^8
V. costi.	-25	8×10^8	1×10^9	1×10^9	1×10^9	5×10^8	4×10^8
	-60	8×10^8	9×10^8	6×10^8	4×10^8	4×10^7	1×10^7
L. bifidus	-25	1×10^9	3×10^8	2×10^8	4×10^7	6×10^7	8×10^7
	-60	1×10^9	1×10^9	1×10^9	8×10^8	9×10^8	1×10^9
L. fermenti	-25	4×10^8	5×10^8	2×10^8	2×10^8	1×10^8	9×10^7
	-60	4×10^8	8×10^8	5×10^8	4×10^8	3×10^8	5×10^8
Ser. marces.	-25	6×10^9	6×10^9	4×10^9	6×10^9	4×10^9	4×10^9
	-60	6×10^9	5×10^9	4×10^9	6×10^9	4×10^9	5×10^9

References

1. Lion, M.B. and Bergmann, E.D. The effect of oxygen on freeze-dried Escherichia coli. J. Gen. Microbiol. 24: 191-200. 1961.
2. Weiser, R.S. and Hennum, L.A. Studies on the death of bacteria by drying. J. Bact. 54: 17-18. 1947.
3. Proom, H. and Hemmons, L.M. The drying and preservation of bacterial cultures. J. Gen. Microbiol. 3: 7-18. 1949.
4. Heckly, R.J., Anderson, A.W. and Rockenmacher, M. Lyophilization of Pasteurella pestis. Appl. Microbiol. 6: 255-261. 1958.
5. Heckly, R.J., Faunce, K., Jr. and Elberg, S.S. Lyophilization of Brucella melitensis. Appl. Microbiol. 8: 52-54. 1960.
6. Harrison, A.P., Jr. Causes of death of bacteria in frozen suspensions. Antonie van Leeuwenhoek 22: 407-418. 1956.
7. Clement, M.T. Effects of freezing, freeze-drying, and storage in the freeze-dried and frozen state on viability of Escherichia coli cells. Can. J. Microbiol. 7: 99-106. 1961.

THE STORAGE OF MYCOBACTERIA IN THE DRIED STATE:
THE EFFECT OF THE CULTURE MEDIUM

by P. W. Muggleton
Glaxo Laboratories Ltd.
Greenford, Middlesex, England

Introduction

The storage of cultures of mycobacteria in the dried state
has been extensively studied. In particular, many studies have been
made with BCG intended for freeze-dried vaccine. Despite the belief that the mycobacteria in general are highly resistant to desiccation, a very considerable loss of viability is almost always
observed when they are freeze-dried. Moreover, once dried, the
cultures tend to be less stable than would be expected, particularly
at temperatures above 25°C.

Almost all attempts to improve this position have concentrated on the freeze-drying medium (1, 2, 3), claims being made that
sodium glutamate, among other things, when added to the drying
medium increases the stability of the cultures. Greaves (4) in
studies with other organisms has shown that the addition of sodium
glutamate very considerably enhanced survival on freeze-drying,
and Scott (5) has shown the rationale behind the use of this substance. He showed, in studies with Salmonella newport, that in the
dried state a reaction occurs between available carbonyl groups
and amino end groups of cellular proteins. The resulting "browning
reaction" compounds are toxic and result in the death of the cells.
The reaction is somehow inhibited by the addition of sodium glutamate which acts as a "carbonyl buffer".

Despite this finding, in our laboratory the addition of sodium
glutamate to the drying medium for BCG did not result in any spec-
tacular increase in survival or subsequent stability. We therefore
decided to look further for the source of carbonyl substance which
might be damaging the cells.

Most cultures of mycobacteria are grown on a culture medium
containing glycerol. They utilise this substance as a source of carbon and produce metabolites (including "glyceraldehyde") which are
potent sources of carbonyl groups. It was therefore decided to grow
BCG on a medium without glycerol and study its stability on drying
and storage.

Experimental

A medium was devised in which the Copenhagen substrain of
BCG would grow satisfactorily without glycerol. Control cultures
were grown in Sauton's medium (containing glycerol). Triton WR-
1339 (0.025%) was added to both media to obtain "deep growth".
The formulae of the two media are shown in Table I.

TABLE I

Glycerol-free medium for the growth of BCG

Sauton's medium			Glycerol-free medium		
L asparagine	4.0	gm	L asparagine	4.0	gm
Ferric ammonium			Ferric ammonium		
citrate	0.05	gm	citrate	0.05	gm
Citric acid	2.0	gm	Monosodium glutamate	4.0	gm
Glycerol	40	ml	Bacto casitone	1.0	gm
K_2HPO_4	0.5	gm	L glutamine	4.0	gm
$MgSO_4$	0.5	gm	KH_2PO_4	1.0	gm
Triton WR1339	0.25	gm	Na_2HPO_4	2.5	gm
Water to	1 litre		$CaCl_2$	0.001	gm
			$CuSO_4$	0.0005	gm
			$ZnSO_4$	0.0005	gm
			Triton WR1339	0.25	gm
			Water to	1 litre	

Cultures of both BCG and M. tuberculosis var. hominis (strain 666) were grown in the two media (100 ml amounts in 1 litre bottles) for 14 days at 37°C. The cells were collected in the centrifuge and resuspended in a freeze-drying medium consisting of dextran (8.3%) and glucose (7.5%) in distilled water. 0.5 ml amounts were filled into ampoules and freeze-dried, in an apparatus which has been described previously (1), at a constant temperature of -28°C.

Viability counts were carried out on samples of the cultures before and after drying using a modification of the method of Miles et al. (6). The results with BCG, which were similar to those obtained with M. tuberculosis (hominis) are summarised in Table II.

TABLE II

Survival on freeze-drying of BCG
grown in glycerol-free medium

Medium	Percentage survival			
	Batch 1	Batch 2	Batch 3	Average
Glycerol-free	53	58	60	57
Control (Sauton's)	25	36	26	29

Ampoules of BCG culture from a batch prepared in glycerol-free medium and a control batch from Sauton's medium were incubated at 37°C. At intervals of time up to 6 months, 6 ampoules from each lot were tested for viability. The results are summarised in Table III.

TABLE III

Survival of BCG freeze-dried
cultures on storage at 37°C

Medium in which grown	Average viability count (as percentage of starting value) at (months):				
	0	1	2	4	6
Glycerol-free	100	86	44	23	11
Control (Sauton's)	100	16	4	0.6	0.01

For experimental work it proved inconvenient to carry out such long term comparative tests and an "accelerated" comparison of stability at 70°C was made. The ampoules were immersed in a waterbath. The results are summarised in Table IV.

TABLE IV

Survival of BCG freeze-dried
cultures on storage at 70°C

Medium in which grown	Average viability count (as percentage of starting value) at (hours):				
	0	1/2	1	3	6
Glycerol-free	100	94	64	-	46
Glycerol-free*	100	88	77	70	61
Control (Sauton's)	100	0.7	0.5	0.1	0.07

*0.02% glycerol added to the freeze-drying medium

Discussion

Cultures of BCG or M. tuberculosis (hominis), grown in a medium without glycerol present, survive the process of freeze-drying better (Table II) and are subsequently more stable on storage at 37°C (Table III) or at 70°C (Table IV). The addition of glycerol to the freeze-drying medium did not decrease the heat resistance of BCG cells grown in the absence of glycerol. This adds support to the hypothesis that it is some metabolite of glycerol (probably glyceraldehyde) which enters into an injurious reaction and not glycerol itself. It is suggested that cultures of mycobacteria which are to be freeze-dried should be grown in a medium which does not contain either glycerol or any other substance which may be metabolised to give rise to carbonyl containing compounds.

References

1. Muggleton, P. W. Freeze-drying of bacteria with special reference to BCG. In Recent research in freezing and drying. Blackwell, Oxford. p. 229. 1960.

2. Miller, R. and Goodner, K. Studies on stability of lyophilized
   BCG vaccine. Yale J. Biol. Med. 25: 262. 1953.

3. Cho, C. and Obayashi, Y. Effect of adjuvant on preservability
   of dried BCG vaccine at 37°C. Bull. Wld. Hlth. Org. 14: 657.
   1956.

4. Greaves, R.I.N. Some factors which influence the stability of
   freeze-dried cultures. In Recent research in freezing and
   drying. Blackwell, Oxford. p. 203. 1960.

5. Scott, W.J. A mechanism causing death during storage of
   dried micro-organisms. In Recent research in freezing and
   drying. Blackwell, Oxford. p. 188. 1960.

6. Miles, A.A., Misra, S.S. and Irwin, J.O. Estimation of bac-
   tericidal power of blood. J. Hyg. (Camb.). 38: 732. 1938.

THE INFLUENCE OF FREEZE-DRYING ON THE
METABOLISM OF MYCOBACTERIA

by R. Bonicke
Institute for Experimental Biology and Medicine
Borstell, Germany

For some years bacteriologists have preferred enzymatic properties for the identification and classification of mycobacteria. Therefore it is important and necessary to know whether or not the enzymatic equipment of mycobacteria shows any change after storage in the freeze-dried state. Thus we have studied this problem.

Of the numerous enzymes we have studied, I have selected those which have a special importance for the classification of mycobacteria.

1. The acylamidatic activity of mycobacteria
   before and after storage.

In our work on the distribution of amidases within the genus Mycobaterium it became clear that various "simple" amidases were confined to certain species. For example, benzamidase, an enzyme which catalyses the deamidation of benzamide, occurs in M. smegmatis only, and it is present in all strains of this species. Succinamidase shows the same specificity to M. smegmatis and it is found in no other species. M. minetti (= M. fortuitum) is characterized by being able to break down allantoin with the formation of glyoxylic acid, carbon dioxide and ammonia. M. thamnopheos and some strains of M. smegmatis are the only other species with this power. In this so-called amide series ("Amid-Reihe") the various mycobacteria show a characteristic amidase spectrum, so that identification of a strain can be effected in a relatively short time by relatively simple means. The possibilities of this method in the classification of rapid growing and atypical mycobacteria is demonstrated in figures 1-7. Mycobacterium smegmatis liberated ammonia (detected by the method of Russell) from benzamide, urea, isonicotinamide, nicotinamide, pyrazinamide and succinamide but not from malonamide (Fig. 1). Allantoinase, salicylamidase and acetamidase are also present in M. smegmatis but their activity is low and incubation longer than 4 hours is needed to demonstrate them.

Mycobacterium phlei has a much narrower amidase spectrum (Fig. 2). This species possesses urease, nicotinamidase and pyrazinamidase. After prolonged incubation (12-14 hours) a positive reaction also appears with acetamide.

Mycobacterium fortuitum (Fig. 3) is distinguished from other rapidly growing mycobacteria by its high acetamidase activity, and

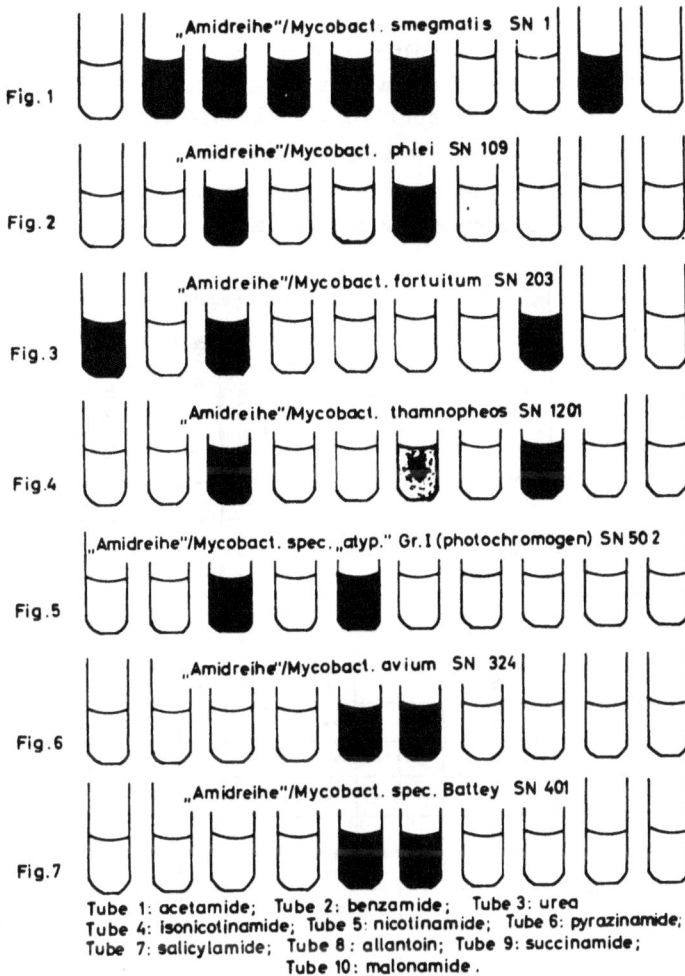

Tube 1: acetamide; Tube 2: benzamide; Tube 3: urea
Tube 4: isonicotinamide; Tube 5: nicotinamide; Tube 6: pyrazinamide;
Tube 7: salicylamide; Tube 8: allantoin; Tube 9: succinamide;
Tube 10: malonamide.

Fig. 1-7. Amidase spectrum of rapid growing and atypical
          mycobacteria.

by its power to break down urea and allantoin. After long incubation nicotinamide and pyrazinamide are also split, but no other
amides in the series are acted upon. A very similar spectrum is
shown by M. thamnopheos (Fig. 4). Photochromogenic mycobacteria of the species M. kansasii, possess only urease and nicotinamidase (Fig. 5). Avian strains (M. avium) can break down
nicotinamide and pyrazinamide only. The origin of the strain is
without influence. Strains of the nonphotochromogenic group of
atypical mycobacteria (Battey strains) possess the same amidase
spectrum. Because of the importance of the amidatic properties

of mycobacteria for their classification, we have examined these properties before and after storage in the freeze-dried state. The freeze-dried strains were recultivated on Lowenstein-Jensen medium. As Table I shows, storage in the freeze-dried state had no influence on the amidatic spectrum of the strains tested.

TABLE I

Acylamidatic spectrum of some mycobacteria
before and after storage

Mycobacterium species	strain	time of investigation *	Enzymatic hydrolisation of										time of reaction in hours
			Acetamide	Benzamide	Urea	Jsonicotinamide	Nicotinamide	Pyrazinamide	Salicylamide	Allantoin	Succinamide	Malonamide	
M. smegmatis	SN 2	a	(+)	+++	+++	++	+++	+++	-	-	+++	-	4
		b	(+)	+++	+++	++	+++	+++	-	-	+++	-	4
M. fortuitum	SN 203	a	+++	-	+++	-	-	-	-	+	-	-	4
		b	+++	-	+++	-	-	-	-	+	-	-	4
M. thamnopheos	SN 1201	a	-	-	+++	-	-	+	-	++	-	-	12
		b	-	-	+++	-	-	+	-	++	-	-	12
M. phlei	SN 109	a	-	-	+++	-	+	++	-	-	-	-	4
		b	-	-	+++	-	+	++	-	-	-	-	4
M. kansasii	SN 501	a	-	-	+++	-	+++	-	-	-	-	-	22
		b	-	-	+++	-	+++	-	-	-	-	-	22
M. aquae	SN 632	a	-	-	+++	-	-	-	-	-	-	-	22
		b	-	-	+++	-	-	-	-	-	-	-	22
M. avium	SN 327	a	-	-	-	-	++	++	-	-	-	-	22
		b	-	-	-	-	++	++	-	-	-	-	22
M. avium (Battey)	SN 404	a	-	-	-	-	++	++	-	-	-	-	22
		b	-	-	-	-	++	++	-	-	-	-	22
M. ulcerans	SN 421	a	-	-	-	-	++	++	-	-	-	-	22
		b	-	-	-	-	++	++	-	-	-	-	22

2. The influence of storage on nicotinic acid metabolism.
   (a) Biosynthesis of nicotinic acid.
      Pope and Smith (1946) were the first to show that human and bovin tubercle bacilli can be differentiated by quantitative differences in vitamin-B-complex production. They found in cultures of the human strain H37Rv higher vitamin concentrations than in bovine strains. Konno and his collaborators investigated a great

number of strains of different mycobacterium species and found
that only human tubercle bacilli give a positive niacin reaction.
All others, bovine, avian, saprophytic and atypical mycobacteria,
give a negative reaction. Since the niacin test is widely used to
differentiate mycobacterium strains, we have studied the ability
of several strains of tubercle bacilli to produce nicotinic acid
after storage in the freeze-dried state. In figure 8 we find the results of these investigations for the human strain H37Ra and the

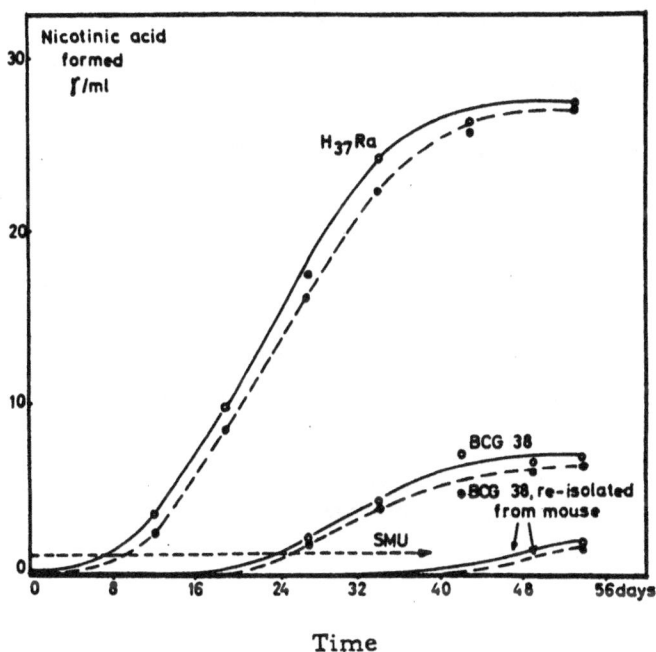

Time

Fig. 8. Formation of nicotinic acid by M. tuberculosis H37 and
M. bovis BCG strain 38 in Lockemann's synthetic medium.
SMU = sensitivity of colorimetric method
o ——— o before storage, ●— —● after storage

bovine strain BCG 38. H37Ra produces a very high concentration
of nicotinic acid, BCG 38 only a small amount; after passage
through mouse the amount is so small, that only after a long time
of cultivation can we determine nicotinic acid formed. Storage in
the freeze-dried state is without influence on nicotinic acid formation.

(b) Deamidation of nicotinamide.

Nicotinamide is an acylamidase which catalyses deamidation
of nicotinamide to nicotinic acid with the liberation of ammonia.
The most important result of our experiments (Lisboa, Bonicke
and Lisboa) about the existence of nicotinamidase in mycobacteria

is that strains of the human type show very high activity in comparison with bovine strains which in general do not show any enzyme activity or only on the condition that we use a large amount of bacteria and a very long reaction time. For this reason differentiation of both types of tubercle bacilli by this test is possible. Konno, Nagayama and Oka confirmed the differences in nicotinamidase activity in human and bovine strains, but their opinion, that only bovine strains have this low nicotinamidase activity, is not true because M. thamnopheos and some strains of M. aquae also have very low nicotinamidase activity. As figure 9 shows,

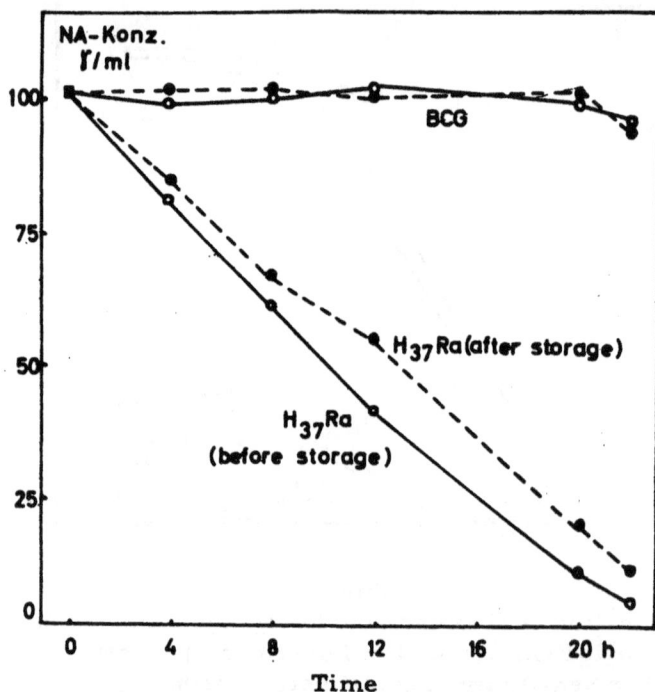

Fig. 9. Deamidation of nicotinamide by M. tuberculosis. H37 and M. bovis BCG.

freeze-drying has no influence on the nicotinamidase activity of mycobacteria. H37Ra has the same high enzyme activity after 4 years storage in the freeze-dried state and BCG 38 retains its low nicotinamidase activity.

(c) Degradation of nicotinic acid.

Whereas M. tuberculosis is the only species forming nicotinic acid, M. smegmatis is the only species of the genus which breaks down nicotinic acid. Figure 10 shows the results of our experiments on the influence of freeze-drying on the enzymatic degradation of nicotinic acid by M. smegmatis SN 2. There exists

no significant difference in activity of this enzyme system, if we
test the strain before or after storage in the freeze-dried state.
This holds for adapted cells as well as for non-adapted cells.

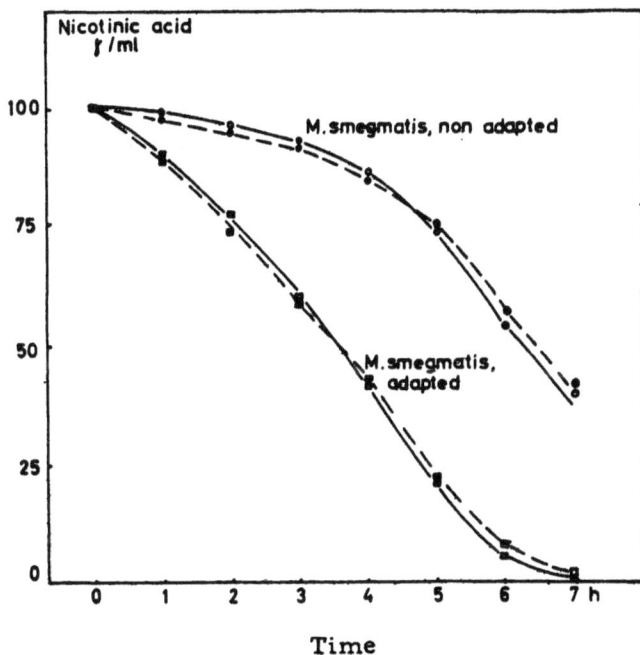

Time

Fig. 10. Degradation of nicotinic acid by M. smegmatis.
o——o before storage, o --- o after storage.

Literature

Bonicke, R. L Zbl. Bakter. I. Orig. 178: 186. 1960; 178: 209. 1960;
        178: 223. 1960; German Medical Monthly 5: 232. 1960.
        Jahresber. Borstel 5. Bd. p. 7-87. Springer-Verlag 1961;
        4. Bd. p. 43-172. 1956/57.
Bonicke, R. and B. P. Lisboa. Tuberk.-Arzt 13: 375. 1959; Zbl.
        Bakter. I. Orig. 175: 403. 1959; Tuberk.-Arzt 12: 380. 1958.
Konno, K. Science 124: 985. 1956.
Konno, K., R. Kurzmann and K. T. Bird. Amer. Rev. Tuberc. 75:
        529. 1957.
Lisboa, B. P. Naturwiss. 44: 617. 1957.
Pope, H. and D. T. Smith. Amer. Rev. Tuberc. 54: 559. 1946.
Russell, J.A. J. Biol. Chem. 156: 457. 1944.

LIST OF DELEGATES

LIST OF DELEGATES

AUSTRALIA

Sherwood, I.R., Sydney
Skerman, V.B.D., Brisbane

BELGIUM

van Dijck, P., Leuven
Hennebert, G.L., Heverlee
Janssens, P.G., Antwerp

BRAZIL

Baracchini, O., Ribeirao Preto
Oliveira de Almeida, Ribeirao Preto

CZECHOSLOVAKIA

Blaskovic, P., Bratislava
Lysenko, O., Prague
Mallek, I., Prague

DENMARK

Halkier, S. B., Copenhagen
Lorck, H., Ballerup
Lund, A., Copenhagen
von Magnus, H., Copenhagen
Rosendal, K., Copenhagen

FRANCE

Auclair, J.E., Jouy-en-Josas
Courtieu, A.L., Lyon
Delaporte, B.L.S., Paris
Enjalbert, L., Toulouse
Nicot, J., Paris
Wahl, R., Paris

GERMANY

Bonicke, R., Borstel über Bad Oldesloe
Giesbrecht, P., Berlin
Glathe, H., Giessen
Hofmann, S., Berlin
Seeliger, H.P.R., Venusberg
Staib, F., Wurzburg
Wagner, B., Berlin
Wundt, W., Tübingen

GREAT BRITAIN

Anderson, E.S., London
Asheshov, E.A., London
Austwick, P.K.C., Weybridge
Barnes, E.M., Cambridge
Bradstreet, C.M.P., London
Brady, B.L., Nutfield
Carpenter, P., London
Collins, V.G., Ambleside
Cooper, K.E., Bristol
Cowan, S.T., London
Dawson, E.R., London
Elphick, J.J., Kew
George, E.A., Cambridge
Hobbs, G., Aberdeen
Marshall, M.J., Harrow
Miall, L.M., Richborough
Miles, A.A., London
Muggleton, P.W., Greenford
Nutman, P.S., Harpenden
Proom, H., Beckenham
Rhodes, M.E., Aberystwyth
Sharpe, M.E., Reading
Shewan, J.M., Aberdeen
Sneath, P.H.A., London
Stewart, I.O., Runcorn
Williams, R.E.O., London
Woodbine, M., Loughborough
Zinnemann, K.S., Leeds

HONG KONG

Stenton, H., Hong Kong

INDIA

Desai, S.C., Bombay
Iyengar, M.R.S., Baroda

IRELAND

Gilliland, R.B., Dublin
Küster, E., Dublin

ISRAEL

Grossowicz, N., Jerusalem
Olitzki, A.L., Jerusalem

ITALY

Babudieri, B., Rome
Hengeller, C., Naples
Silvestri, L. G., Milan
Virgilio, A., Naples

JAPAN

Asai, T , Tokyo
Hasegawa, T., Osaka
Nei, T., Sapporo
Okumura, S., Kawasaki
Takahashi, M , Kawasaki
Terui, G., Osaka

NETHERLANDS

van Beverwijk, A. L , Baarn
Rörsch, A., Rÿswÿk
van der Waard, W F., Delft
Wikén, T. O., Delft
Wolff, H. L., Leiden
Wolff, J W., Amsterdam

NEW ZEALAND

Blair, I. D , Canterbury
Miles, J. A. R., Dunedin

NORWAY

Jonsen, J., Oslo
Lahelle, O., Oslo
Lindeberg, G., Vollebekk

PAKISTAN

Chughtai, M. I. D., Lahore
Muhammed, A., Lahore

SPAIN

Holguera, F., Leon
Rejas, F., Leon
Vicente-Jordana, R., Madrid

SWEDEN

Hedén, C. G., Stockholm
Juhlin, I., Malmo
Molin, N., Gothenburg
Norén, B., Lund

Nordberg, B.K., Stockholm
Norkrans, B., Goteborg
Paldrok, H., Stockholm
Tveit, M., Arlov

SWITZERLAND

Cockburn, W.C., Geneva
Hütter, R., Zurich
Nüesch, J., Basel

TAIWAN

Liu, Y.P., Taipei
Tai, F.H., Taipei

U.S.A.

Abrams, A., Washington
Ahearn, D.G., Miami
Alexander, M.T., Washington
Baker, C., Waltham
Barratt, R.W., Hanover
Baugh, R.J., Bloomington
Beckhorn, E. J., Staten Island
Berger, J., Passaic
Bitter, C.R., Boulder
Bloomfield, B.J., New Brunswick
Bradley, S. G., Minneapolis
Braendle, D.H., North Chicago
Brillaud, A.R., Marcus Hook
Brown, W.E., New Brunswick
Buchanan, R.E., Ames
Burkholder, P.R., Palisades
Burns, M.E., Bethesda
Charney, W., Union
Christensen, C.W., Detroit
Clark, W.A., Washington
DeBecze, I. G., Lawrenceburg
Diamond, L. S., Bethesda
Dietz, A., Kalamazoo
Donovick, R., New Brunswick
Eveleigh, D.E., Natick
Fennell, D. I., Madison
Goos, R.D., Bethesda
Gordon, R.E., New Brunswick
Hansen, P.A., College Park
Harrison, A.P., Jr., Nashville

Haynes, W.C., Peoria
Hillegas, A.B., Detroit
Heckly, R. J., Oakland
Herman, L.G., Bethesda
Hugh, R., Washington
Hwang, S., Washington
Kline, I., Washington
Krabek, W.B., East Weymouth
Lessel, E.F., Washington
Levine, N.D., Urbana
Lindberg, R.B., Fort Sam Houston
Lindblom, G.P., Tulsa
Mazur, P., Oak Ridge
Meyers, S. P., Miami
Middaugh, P.R., Muscatine
Moline, S. W., Tonawanda
Ogata, W.N., Hanover
Parker, R.F., Cleveland
Pratt, D.B., Gainesville
Raper, K.B., Madison
Rosenberg, S., Millburn
Rosenman, S. B., Philadelphia
Routien, J. B., Groton
Schocher, A. J., Boonton
Sellars, R.L., Milwaukee
Shahidi, S., Madison
Shannon, J. E., Washington
Simmons, E.G., Natick
Stapley, E.O., Rahway
Starr, M.P., Davis
Starr, R.C., Bloomington
Stevenson, R.E., Silver Spring
Stulberg, C. S., Detroit
Thompson, R.L., Bethesda
Tresner, H.D., Pearl River
Truant, J. P., Birmingham
Updyke, E., Atlanta
Uridil, J. E., St. Joseph
Vakil, J. R., Lincoln
Valladares, Y., Houston
Vavra, J. J., Kalamazoo
Ward, B. Q., Hattiesburg
Winn, J. F., Atlanta
Wolfson, S., Millburn

U.S.S.R.

Krasilnikov, N.A., Moscow
Michoustine, E.N., Moscow
Preobrazenskaja, T.P., Moscow

CANADA

Amies, C.R., Toronto
Blackwood, A.C., Macdonald College
Boylen, J.B., Montreal
Bynoe, E.T., Ottawa
Byrne, J.L., Hull
Carmichael, J.W., Edmonton
Clark, D.S., Ottawa
Clark, W.A., Montreal
Clegg, L.F.L., Edmonton
Clement, M.T., Ottawa
Colwell, R.R., Ottawa
Cook, W.H., Ottawa
Doyle, R.J., Windsor
Elliott, J.A., Ottawa
Elliott, M.E., Ottawa
Farkas-Himsley, H., Toronto
Fukui, S., Ottawa
Gibbons, N.E., Ottawa
Gray, P.J., Montreal
Groves, J.W., Ottawa
Haskins, R.H., Saskatoon
Hauschild, A.H.W., Kingston
Hawirko, R.Z., Winnipeg
Ingram, D.G., Guelph
Katznelson, H., Ottawa
Khairat, O., Winnipeg
Maim, H., Toronto
Magus, M., Toronto
Martin, S.M., Ottawa
McDonald, I.J., Ottawa
Meerovitch, E., Macdonald College
Merger, C.E., Toronto
Morgan, J.F., Saskatoon
Murray, E.G.D., London
Nagler, F.P., Ottawa
Nogrady, G., Montreal
Prakash, A., St. Andrews
Prytula, A., Toronto
Quadling, C., Ottawa

Rouatt, J. W., Ottawa
Simpson, F. J., Saskatoon
Timonin, M. I., Saskatoon
Ueno, T., Ottawa
Vezina, C., Montreal
Webb, S. J., Saskatoon
Wellman, A. M., London
Westlake, D. W. S., Saskatoon

www.ingramcontent.com/pod-product-compliance
Lightning Source LLC
Chambersburg PA
CBHW051753200326
41597CB00025B/4537